Rise of the Badger
and the Great Shrubbery

Rise of the Badger
and the Great Shrubbery

Compiled by Peter Lay

What phoenix will arise from the ashes of the human race?

Black Eyes Publishing UK

Rise of the Badger and the Great Shrubbery
© Black Eyes Publishing UK, 2025

Published 2025
Black Eyes Publishing UK
Gloucester GL1 3ET (UK)

www.blackeyespublishinguk.co.uk

Paperback - ISBN: 978-1-913195-38-0

The content this anthology remains within the copyright of the individual writers. They have asserted their moral right under the Copyright, Designs and Patents Act, 1988, to be identified as the authors of their work.

All Rights reserved. No part of this publication may be reproduced, copied, stored in a retrieval system, or transmitted, in any form or by any means, without the prior written consent of the copyright holder(s), nor be otherwise circulated in any form of binding or cover other than that in which it is published and without a similar condition being imposed on the subsequent purchaser.

No part of this book may be used or reproduced in any manner for the purpose of training artificial intelligence technologies or systems. In accordance with Article 4(3) of the Digital Single Market Directive 2019/790, the author and publisher expressly reserve this work from the text and data mining exception.

A CIP catalogue record for this title is available from the British Library.

Compiled by: Peter Lay

Cover design: Jason Conway, The Daydream Academy.
                                  www.thedaydreamacademy.com

Back Cover Artwork: Neil Windsor,
                                          Badgers on Zebra Crossing in post-apocalyptic Leeds

Let the mycelium inherit the earth,
silently fruiting in damp places
a unity we never understood.

Rhian Thomas
(from, 'the Grandmother Hypothesis')

We may be the only species to choose extinction because
it isn't cost effective to save ourselves.

Rebecca Clifford
(from, 'My mother was a whizz in the kitchen')

PREFACE

The idea for this Anthology was born when I heard, yet again, someone talking about 'saving the planet'. It occurred to me that it was humanity that needed to be saved; what they were actually talking about was the demise of the human race. Given time, the chances are, the planet will be fine. The planet will survive the climate crisis and the endless wars; people won't. Even these terms; 'Planet', 'Earth', 'Gaia' etc. are all human titles. No humanity; no names; no Earth. When we cease to be, our language goes as well. I doubt the dinosaurs created a name for the place they inhabited.

Are we already too late to prevent it? Probably!
Should we put the proverbial paper bag over our heads, get under a table and just fade away?
No! It's not all doom and gloom. What do we do now? Each of us should strive to do and be the best we can...Live each of our individual lives towards a glorious zenith.

Around a decade or so, at the foot of the Cheese-rolling hill on the A46 between Brockworth and Painswick in Gloucestershire UK, there were two car parks. The first, near a quarry, was closed due to subsidence. Now, as you drive past you might never know that once there was a place to park. The whole area is filled with dense shrubbery and trees are taking over, it has become part of the wooded slope of the hill. It may also surprise some people to know that the top land predator in the United Kingdom is the Badger, which like humans are omnivores, therefore adaptable foodwise. Hence the title came to me. **'Rise of the Badger and the Great Shrubbery'**.

This will be a Climate Crisis Anthology with a difference, more a Climate Disaster! Here we will assume that the worst has happened; the Human Race has destroyed itself. What comes next?

What kind of phoenix will arise from the ashes of our human world?

Peter Lay
Black Eyes Publishing UK

CONTENTS

7	PREFACE
13	INTRODUCTION
17	*Untitled One (drawing)* ~ GURINDER SINGH KALSI
19	*On a Different Page (poem)* ~ ADAM HOROVITZ
21	*These Facilities Are Not Exceptional (poem)* ~ RHIANNA LEVI
22	*After the concert (poem)* ~ ROGER TURNER
23	*Pellitory of the Wall (poem)* ~ ROGER TURNER
25	*From the Stump, a Forest Rose (poem)* ~ EMMA DAVIDSON
29	*Ravens (watercolour, mixed media)* ~ EMMA BURLEIGH
31	*The Rise of the Crows (poem)* ~ MICHAEL DAVIES
33	*On not and always existing (flash-fiction)* ~ STEPHANIE CARTY
34	*The Silence That Remained (poem)* ~ MICHAEL DAVIES
36	*Keep Off The Grass (poem)* ~ CATHERINE MARINA
37	*So here we are (poem)* ~ ANDREA TILLMANNS
39	*Badgers on Coopers Hill (mixed media collage)* ~ STEPHANIE CARTY
41	*The Commons (poem)* ~ RHIAN THOMAS
42	*Elegy for a Parasite (poem)* ~ SIMON ALDERWICK
43	*Badger (poem)* ~ DAVID THOMPSON
44	*The Haunted Library (poem)* ~ BRUCE McRAE
46	*Cats and Dogs (flash-fiction)* ~ FAITH EAGLES
48	*When We're Gone (poem)* ~ STEPHEN CHAPPELL
50	*Sacred Disc (poem)* ~ DJ TYRER
51	*Dead Sands (poem)* ~ DJ TYRER
52	*Tardigrades - for the love of tiny things (poem)* ~ JOSEPHINE LAY
55	*Lion (acrylic)* ~ MARY PEARCE
57	APEX PREDATOR ~ LION
59	*The Birds are Hungry (poem)* ~ NATASHA GAUTHIER
60	*A Dog-Eat-Dog World? (Flash Fiction)* ~ ALEX PERRY
63	*Cult of the Badger (painting)* ~ MORGAN RYE
65	*Rise of the Badger (poem)* ~ PETER LAY

67	*Aftersong (poem)* ~ CHRISTAN WARD
68	*Songs of the Past (flash-fiction)* ~ CATH HUMPHRIS
73	*Badger in the Gloaming (watercolour, mixed media)* ~ EMMA BURLEIGH
75	*Love song of the badger (poem)* ~ STEWART CARSWELL
76	*Time Comes Undone (poem)* ~ ELLIE LI
77	*Above the Burrow (flash-fiction)* ~ TOM EDMONDS
81	*Post Apocalyptical Zebra Crossing (artwork)* ~ NEIL WINDSOR
83	*Post Apocalypse Disco Badgers (poem)* ~ NEIL WINDSOR
85	*Post Apocalyptical Disco Badgers (artwork)* ~ NEIL WINDSOR
87	*Getting rid of old pizza boxes (artwork)* ~ NEIL WINDSOR
89	*The Animal (poem)* ~ JAMES KENNY
90	*When we are Gone (poem)* ~ CHRISTINE GRIFFIN
91	*3rd Time Charmed (flash-fiction)* ~ JENNIFER LAXTON
95	*Old House in Moon Forest (watercolour)* ~ RUTH SCHREIBER
97	*Re-Taken (prose)* ~ LEE McSHANE
98	*Gracie (flash-fiction)* ~ DOUG DEVANEY
101	*Written Fates (poem)* ~ CHARLES CUYANA
102	*The Voice in its Many Mansions (poem)* ~ BRUCE McRAE
105	*Saltwater Crocodile (photo)* ~ GETTY IMAGES FOR UNSPLASH
107	*APEX PREDATOR ~ SALTWATER CROCODILE*
109	*The Smog (a conversation)* ~ CHRISTOPHER T. DABROWSKI
110	*Dragons Sleep with One Eye Open (Tanka)* ~ PETER DEVONALD
111	*The Eyes of The Dragon (poem)* ~ PETER DEVONALD
112	*The Leftovers (poem)* ~ PETER DEVONALD
113	*Nigh (poem)* ~ CHRIS HEMINGWAY
114	*Wegotu (poem)* ~ STEPHEN LITTLEJOHN
117	*Polar Bear (photo)* ~ ALEX ROSE
119	*APEX PREDATOR ~ POLAR BEAR*
121	*The polar bears have moved into the houses (poem)* ~ CAROL SHEPPARD
122	*It is the New Normal (flash-fiction)* ~ SHASHI KADAPA
125	*Untitled Two (drawing)* ~ GURINDER SINGH KALSI
127	*Microbial Soup (poem)* ~ RICHARD CATLIN
129	*Timber Cries (poem)* ~ TREVOR VALENTINE
131	*Repairing the Planet (painting)* ~ GURINDER SINGH KALSI

133	*Report 06172 (flash-fiction)* ~ JOHN GRIEVE
134	*Unobserved (poem)* ~ TONY BRADLEY
135	*You Are Here (poem)* ~ AVRIL O'LEARY
137	*Tiger (etching)* ~ HELEN PREDGEN-LAY
139	APEX PREDATOR ~ TIGER
141	*Where The Wild Things Aren't (poem)* ~ ALBY STOCKLEY
142	*The Somewhat Unexpected Rise of the Lemmings (poem)* ~ CHRIS HEMINGWAY
143	*A Parliament of Owls (poem)* ~ GARETH WRITER-DAVIES
144	*Your Other Half (poem)* ~ JANET PENNEY
147	*Orca (photo)* ~ VIDAR NORDLI-MATHISEN
149	APEX PREDATOR ~ KILLER WHALE (ORCA)
151	*The Planet's Doing Just Fine Thank You (poem)* ~ ROSIE BARRETT
152	*This Red Planet (poem)* ~ ADRIAN McROBB
153	*The Remaining Tasks (flash-fiction)* ~ LOUIE CLARK
156	*Today's Lesson is taken from the Book of the Prophet Apocalypse (poem)* ~ DAPHNE MILNE
157	*Homo Sapiens (poem)* ~ DUNCAN FORBES
158	*Extinction... (poem)* ~ ADRIAN McROBB
159	*Man, Your Victories are in Vain (poem)* ~ IBRAHIM HONJO
160	*Crumbs of earth (poem)* ~ MATHIAS JANSSON
161	*Global Mourning (prose)* ~ DUNCAN FORBES
162	*Afterwards, at Lyme Bay (poem)* ~ ALAN MANSELL
163	*RAT 2.0 (prose)* ~ MARIA DE STEFANO
165	*Dear Former Tennant (poem)* ~ NIGEL KENT
167	*Badger (print)* ~ EMMA BURLEIGH
169	ACKNOWLEDGEMENTS
171	BIOGRAPHIES

INTRODUCTION

Within these pages we will assume that the human race has failed to protect itself and has finally become extinct. What happens next on this small blue planet that floats in a minor solar system within the vastness of space?

From the first illustration entitled, 'Untitled One', we are taken into the unknown with some humour. Is it a plant or strange animal? Is it aquatic or terrestrial? Is it the sunrising or a weird flower, or is it a jester of life laughing at us who are extinct?

Black Eyes had many submissions which were erudite and powerful but which still held elements of human survival within them. Sadly, for the purposes of this Anthology as set out by Peter Lay in his preface, they had to be rejected, because here humanity has ceased to exist. It appears that for some people, even highly creative thinkers, this existential crisis of our species is difficult to conceive. But as you, the reader, move through this collection of poetry, prose and art it is essential that you keep this proposed fact in mind. Some of the pieces are factual, others a comic take on a non-human world, a few are completely off the wall and make the reader stand on their head in order to view what might be totally incomprehensible. The whole Anthology should be an entertaining read; we want you to enjoy it; but there is, obviously, a serious undercurrent of potential or even probable reality highlighted here.

Peter envisaged this project in the hope, possibly a vain hope, that it may make a tiny difference. That those involved in its creation and those who read it might take on board the urgency of our potential demise if we do not make greater efforts to conserve and adapt to the changes in our environment that are occurring at an alarming rate.

The importance of this Anthology has become increasingly apparent; this summer of 2025 has seen extreme temperatures across Europe. Even as I write this introduction, the news is full of images of forest fires raging in Spain, Portugal,

Greece and Albania. Scandinavian countries haven't escaped the hot dry conditions; temperatures rose above 30 degrees in Norway and Finland. Many US states have experienced extreme weather this year with fierce storms, torrential rain and flash floods. There can be little doubt that Global Warming is no longer just a theory. Yet still humanity appears to be unable to work together to minimise the worst that climate change can cause. It prefers to focus on war; as in Ukraine, in Gaza, in Syria and in the Sudan. Vast sums are spent on the arms industry, and the threat of nuclear bombardment is still very real.

Prevention is possible, we are innovative and flexible creatures, but if greed and the pursuit of personal gain and happiness take precedent; if we continue to bury consciousness and concern in the digital sand of screen time; if territorial disputes and wars are more important to us than peace and working in unity, then we are probably doomed and perhaps rightly so. Hence the quote from a poem by Rebbecca Clifford featured at the front of this volume:

> *We may be the only species to choose extinction*
> *because it isn't cost effective to save ourselves.*

Here is sobering thought: if we humans die out, will our God or gods perish with us? Will they be no more than hieroglyphs on a temple wall? All the human pain, tragedy and suffering that has taken place through religious wars and conflict will be as nothing; blown away in the wind of our demise. Only the spirits that inhabit the fauna and flora of this planet might remain.

Josephine Lay – August 2025

* *'My mother was a whizz in the kitchen'* by Rebecca Clifford is not in this anthology, as it is not post-humanity. However, it will be featured in the 2026 Black Eyes anthology, **'Obsessed by Crows'** which will have a loose relationship theme.

GURINDER SINGH KALSI

Untitled One

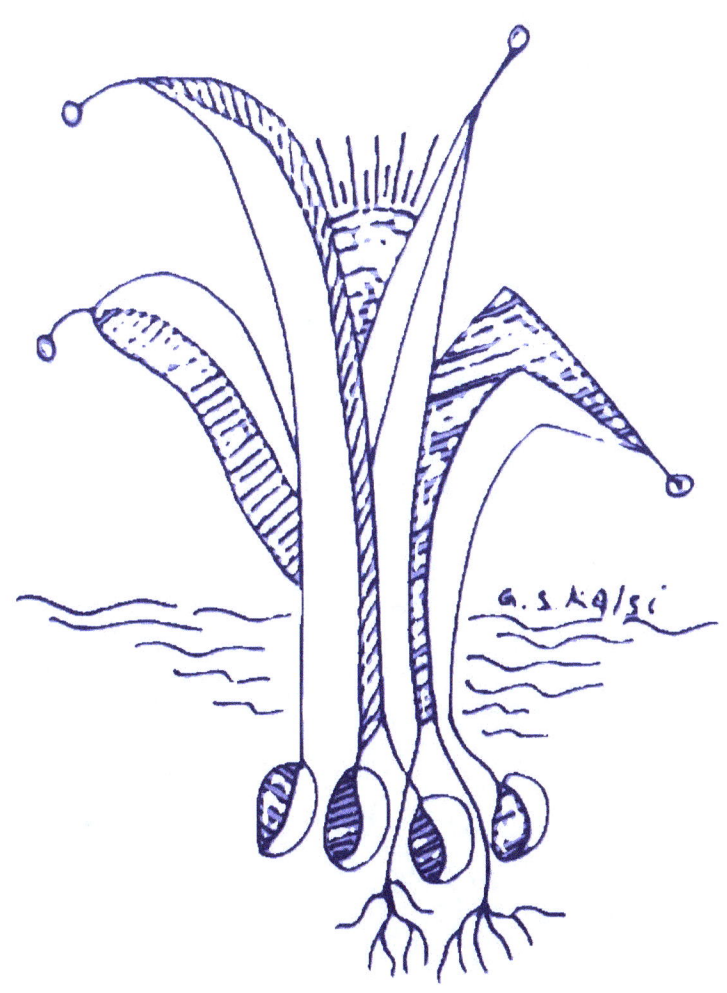

ADAM HOROVITZ

On a Different Page

Someone broke the index
of our existence, stripped
away references for fire lighters,
buried the gutted books of life
in a pit beneath a motorway.

It was never clear who.
The world slipped slowly out of joint
around us. Bees ceased
singing their ancient winter song of warmth
when we passed the hives.

I learned to sympathise
with clip-winged queens
lost in the grass,
unable to lead the swarm,
stripped of leadership, survival, flight.

Nothing worked as it should;
trapped in a spiral of panic,
all our fragmented
moments span into the gutter
like autumn leaves.

We tried to pollinate
our memory with feathers
as the crop of thought
began to fail, but frost
ate the blossom on every tree.

Home sidled out of focus
so we travelled
from house to indistinct house,

trying keys in any door
that looked as if it might be ours.

The people we assumed
were listening within
made no noise,
however much we called
for their attention.

They're on a different page you said
as rusting engines
cropped the dark turf
and turrets flew
out of the flesh of birds.

Finally everything we understood ceased moving.
You faded from reference
just as a hum of new music,
a binding heartbeat of numbers,
swelled beneath broken fragments of Tarmac.

In whispering fields,
the processes of growth
opened up a different book,
its index born of grass
and trees and wild wheat.

Only a few bent leaves
to mark my passage through this wilderness
should someone find them
who can understand. Nowhere else
a reference to the genus of man.

RHIANNA LEVI

These Facilities Are Not Exceptional:

No, we shall not be memorable or be hailed in choral luminosity or verse. Not for our riches, attraction, cognition.

We will be thought of for our neglectful songs, our drive to make the world burn in our frozen ready meals. Squeezing nothing but ourselves. Squirting petrol in delirious directions to be someone's else's complication.

The horned beasts of vampiric ancestry are now the only ones that roam. But even their thrones crumble so comfortably, materials meeting fatalities willingly.

They crunch on nothing but iron gun shells and plastic wrappers that drowned in rivers which once swam. Slurping their own ash in what were once squash beakers.

These facilities are not exceptional, neither are the inhuman beasts that we came to be.

ROGER TURNER

After the concert

At last the passion and the tumult are finished,
the black notes and the white are lying still.
The pianist has gone, and also the audience,
and all is silent in the splendid hall.

Some men who are not famous
come quietly and, without emotion,
lower the curving silhouetted lid -
black as a night sky that has no stars.

The noise, the excitement, the joy, the grief,
all will be forgotten, will fade beyond recall,
when galaxies of stars, suns, planet Earth,
go down into the infinitely dark,

and time that made the heart beat
and moved the hands, gone too.
The music of the sunlight and the starlight
will be over, and the concert done.

ROGER TURNER

Pellitory of the Wall

You are not beautiful at all -
a dusty, harmless, dull green creature
nestling quietly in the corner -
Pellitory of the Wall.

And yet you seem to be a small,
almost-unnoticed messenger,
a step into a leafy future,
Pellitory of the Wall.

No one sees your soft seeds fall;
your tiny stems, as they take root,
are almost trodden underfoot
between the pavement and the wall,

Fleabane of Canada, lank and tall
ubiquitous but alien,
is usually your companion -
Pellitory of the Wall.

But if we do no work at all,
forget all tidying and weeding,
then, on the summer wind self-seeding,
comes Pellitory of the Wall,

with Willowherb and Pimpernel,
Scented Mayweed, grey Fat Hen -
Nature's army flying in,
seeking to recover all.

Day after day they grow more tall,
with buddleia, ash and golden rod,

until the town becomes a wood
and ivy covers every wall.

But you are first, though low and small,
to soften every scar and stain
and make the whole world green again,
Pellitory of the Wall.

'Pellitory of the Wall' won 3rd Prize in the Bridlington Poetry Festival Competition, 2013.

EMMA DAVIDSON

From the Stump, a Forest Rose
For the fallen at Sycamore Gap

It started with a single crack...
this wasn't thunder, but blade,
a whine through dawn mist
where the lone tree stood
rooted in history,
its limbs cradling the sky.

No storm.
No fire.
Just
Man.

Steel teeth through heartwood,
sawdust where shade once danced.
The Sycamore fell,
not with a roar...
but with the silence that comes after a war ends.

Some said it was just a tree.
But we knew.
The moment it hit earth,
the world sighed in grief.
Winds changed direction.
Screens flickered.
The sky dimmed,
and a stillness seeped through steel and stone.

Pipes cracked.
Rails buckled.
Roads broke under their own memory.
The servers hummed one last lullaby

before going dark.
We called it collapse.
Nature called it their return.

From the stump,
green arms reached up.
Shoots like prayers.
Moss crowned the severed bark.
Roots, unshackled, stretching wide,
cracking open the tombs of tarmac.

From gutters sprang brambles,
from satellites, sparrow nests.
Wind sang through shattered glass
while wisteria climbed Parliament's bones.
No kings now.
Only crows.

Badgers walked back into the cities
as if they never left.
No fear in their stride,
only hunger and inheritance.
Shrubs tore through the shopping malls,
spilled through escalators,
as they broke the branded lights
letter by letter.

Oceans now moved differently.
Mountains exhaled.
Fires cooled.
And time slowed.

In place of people,
there were petals.
In place of nations,
there were networks of roots.

No more borders but rivers.
No laws but seasons.

The last poem written by a human
was carved into the stump:
"We had our chance."

And from it,
ferns unfurled.
A forest rose.
Not in rage,
but in a rhythm.
Not in vengeance,
but in total balance.

We are the echoes now.
And the earth sings still.

28

EMMA BURLEIGH

Ravens (Watercolour with mixed media)

MICHAEL DAVIES

The Rise of the Crows

They came as dusk swallowed the land,
on wings like ink, in murder's band.
No funeral bells, no flags, no cries—
only the black rain of their eyes.

From power lines to castle spires,
they perched like judges on old wires.
The fields were theirs, the skies once more,
each city now an open door.

They watched as factories went cold,
as corpses slumped and flesh grew old.
The crows, they neither wept nor prayed—
but marked the cost of what we made.

Where warplanes rusted in the grass,
and schoolyards held no child to pass,
they built their nests in helmets grey
and tore our pages clean away.

Not just birds, but something wise—
with time and silence in their eyes.
They'd watched us rise, they'd seen us fall,
and now they ruled above it all.

A parliament of feathered spies,
who croaked beneath the deadened skies.
Their laws were simple: "Watch and wait.
Take what's yours, abandon hate."

No need for fire, no need for sword,
their patience was their sharpest word.
They didn't burn our homes or write—
they simply watched us lose the fight.

And as the ivy curled and grew,
the crows in black-winged numbers flew.
The world turned quiet, strange, and wide—
and crows became its only guide.

STEPHANIE CARTY

On not and always existing

The body of Adele Quinn lies curled on the ground at the rear of the garden. The sunflowers she planted to absorb toxic heavy metals from the soil have dipped their heads out of respect, seeds dropping here and there, or scooped out by fluttering goldfinches. Without confinement to a cage in a drawing-room, the birds flap then glide from flower to tree in that way Adele had thought joyful. Several swoop upwards, calling out the news that she is gone; they are all gone. A black cat stretches its paws long in front of it, then sits facing away from the dead woman. Adele's husband had always taken this behaviour to be a sign of disdain but she had been sure that creatures have their own ways of doing things.

Truths dissolve along with Adele's soft tissues. It is no longer September twenty-third but the time of year that a new bumblebee queen knows to forage for pollen from dahlias and ivy. The sky is not blue with purple-red tinges to any eyes that still see it. But she was right in that last moment of life with a hazy half-thought of 'this can't be the end'.

Bacteria busy themselves, feeding and fermenting. Insects obey the odour of death. Fungi within take their task seriously. The soil drinks and is thankful.

Over time – this new and old measure of time – Adele Quinn returns her elements of hydrogen, oxygen, carbon and nitrogen. The soil will hold her chemical imprint. Plants nearby will grow faster with altered leaf colour and fluorescence. She is the vegetation. She is the shade and shine. She is the Big Bang, nuclear fusion, the death-throes of stars, eruptions of volcanic gas, ancient bubbling microbes, every human ancestor. She is the end and the beginning.

MICHAEL DAVIES

The Silence That Remained

No more voices in the morning air,
no more hands to build or care.
The silence crept through cities vast,
a final breath, a final gasp.

The echoes faded, street by street,
where laughter danced and lovers'd meet.
Doors hung open, rusted wide,
the wind alone was left to guide.

Glassless towers touched the sky,
their broken windows blinking eyes.
Concrete jungles cracked and bowed,
ferns erupted through the road.

Once we mapped and claimed the stars,
tamed the sea, and healed our scars.
But pride runs thin when tempers flare,
and we forgot the cost of air.

No bombs remained, no final war,
just sickness scratching at the door.
And when it came, it took us slow,
like frostbite creeping toe by toe.

Now foxes pace the shopping aisles,
where lights still flicker once in a while.
An engine hums, then fades again—
a ghost-machine without a name.

And so the Earth, in weary grace,
reclaims each scar, each human place.
It never hated, only bore
the burden of what came before.

We sang, we fought, we made, we tore—
now silence reigns forevermore.

CATHERINE MARINA

Keep   Off   The   Grass

breaking into a garden through a topiary hedge on rusty bikes the last children
snap branches back   trees spit green at them in spite   they laugh
lift their front wheels higher than the back
                                    and roar like dandelions into the garden

the dust is sherbet   the air settles   their eyes adjust to green and pollen
boiling air washes their faces   it   is   bone meltingly hot

the sky rolls towards them in tides   they eat the flowers   smell the soil
suck water out of buttercup stalks   lie grass deep   let blades spike their eyeballs

heat clings to them in shade   forms snow drops on their bodies
washes away building dust   softens concrete skin

they make Grass Angels   flying creatures pulled up from their roots
straw wings catching fire   feathers dripping candlewax as they ascend
into a hydrochloric sky   dinosaurs becoming birds

ANDREA TILLMANNS

So here we are

Everyone knew about the Big Five,
mass extinctions caused by
catastrophes or transitions
between geochronological periods.

Everyone could have known
that the sixth mass extinction
had already begun.

I had no hope
that people would come
to their senses on their own.
I built a time machine
to see and report,
to shake things up and, at the last moment,
to get my fellow humans
to save themselves.

A thousand years into the future,
then back again—that was my plan.
It was warm there, very green, very stormy,
the war-torn cities deserted.
Many animal species smaller, plants larger
than before, hunters and hunted,
everything new
and still searching
for balance.

So I stayed, the last human
in a new, wonderful,
chaotic, unpredictable world
that was quickly
recovering from us.

STEPHANIE CARTY

Badgers on Coopers Hill (mixed media collage - paper + acrylic)
The famous Cheese Rolling Hill in Gloucestershire

RHIAN THOMAS

The Commons
For Dave Green

Living through landslide season, forewarned
and blindsided, we watch things

disgorge. Neighbours and flags cascading by.
Things we buried turned loose

on rift tides, coffins yawning their debris into the waves.
Murk builds its own currents. We huddle, holding ground,

asking where it ends. Up on the commons
you take your hammer, split stones

to tell us time by a different clock. How this cortex
that we balance on weaves together

a million million breathing fragments coalesced
to erupt orchids. How towering measures, eons deep,

eroded unseen from where we stand.
How oceans have closed before

continental margins slipping alongside, things
evolved in isolation stepping over

into new shared ground. How
racing corridors of hawthorn carry endings

and beginnings in the alchemy
of one heady scent.

SIMON ALDERWICK

Elegy for a Parasite

when the last man on Earth breathes his last breath
trees will sigh in relief—they can grow tall again.

no human alive now has ever seen, will ever see,
trees as tall as they used to grow.

when the last man falls, no one will hear him.
animals will be too busy running free.

you can stop the clocks, stop chasing your tail,
bask on a rock by the sea in the sun.

a glint in the eye of every beast
like the silver lining of every cloud.

now birds will sing as they feast on insects
that can finally breathe clean air again

and the fields will grow wild and the ocean
will dance in the moonlight to celebrate

and whales and dolphins will do backflips when
the last man draws his last pitiful breath.

DAVID THOMPSON

Badger

War, pestilence, famine:
human life expires, bones
of the species eaten clean,
drought-whitened,
Jurassic accretions on
the earth's pitted skin.

In the oceans, octopuses
multiply, confident they're
the life form most likely to
intercommunicate,
most apt to impose the planet's
next dominant community.

On land, why look further
than ants and bees, implacably
social, all for one, one for all?

Yet up in the woods,
their setts uncamouflaged
 in a world where humans
are no longer the enemy,
the badgers lose their caution.
There's no need to sniff
to check what's upwind,
little need to stay nocturnal:
ten million years of dark,
a new day of light.

BRUCE McRAE

The Haunted Library

Books written underwater
or in a great hurry.

Books written while contemplating
the moon's grave underside.

Books written by the left hand,
backwards, with eyes shut.

Written naked, the house on fire.
Written in bed, in the dark,
in a nation other than
the author's origin.

Books about books. About authors.
About authors writing books
and never achieving an audience,
unable to shore the river's banks
against the coming Great Erasure.

Books that fed the furnaces of Carthage.

Books banned and burned
in a climate of heady ignorance.

Books within books, their puzzling plots
a blot on the reader's imagination,
characters based on those neither living nor dead,
their lives, their torments, their loves.
Stories told against a backdrop of history
or pure inventions of a riddled mind,
that no one reads, that no one's heard of,

the library haunted, the librarian run off
with the first man to ever have kissed her.

And this book, its dog-eared author,
his face like a blank page,
his glib asides scrawled in the margins.

FAITH EAGLES

Cats and Dogs

They chased the old enemy through the streets, skipping past trees and leaping over the long grass that pushed through the broken concrete like weak teeth, paws swift over the rubble. Their quarry was fast, and the pace it had set was wearing many of them out. They panted hard, all their muscles straining. But the enemy was too far in front, and the group leader watched it with desperate eyes as it disappeared behind a wall.

Dodger, a large, heavy-furred black dog, lifted his head and stopped to look at the tall structures that towered over him. A long time ago, long before Dodger's entry into the world, an unfamiliar creature roamed these lands in their millions. The dogs just looked upon their great works as they would any other rocky landscape. There were no signs of the previous inhabitants now. A distant memory, passed down from parent to pup. Not even a scent drifting in the air existed. Only the smell of the old enemy accosted the dogs' noses, and they would chase it whenever they could. An instinct as old as time itself.

The lead dog searched around the wall that was attached to nothing, just standing amongst crumpled bricks and dust. A lone sentinel, guarding the entry to enemy territory. He shook himself and let his nose do the work. Smells from all around him swirled over his scent receptors, giving him another view of the world. Dodger could sense the old enemy but could not see them. His wet nose twitched, and he turned his greying muzzle to the wind. Like isolating a colour from a rainbow, he identified the scent they had been pursuing from the myriad of others. It followed a path over the rubble and up some steps into a building. Motioning with a tilt of his head to the rest of the pack, he led them past the wall towards it. Dogs of various colours and sizes trotted along, stopping every so often to sniff the cracked ground. The steps led nowhere. A hole in the ceiling was the only entryway, and it was too high for any of them to jump. Some dogs whined sorrowfully. They were hungry, stomachs pained, and prey in the area had been light on the ground. They would have to abandon their quarry and search further afield.

Dodger's ears pricked up at a sound far above them. A low call at first that rose to a screech, that penetrated his ears and made them itch. This pattern repeated several times. Similar cries answered it in the distance, making the pack restless. They whined and growled.

The lead dog searched for a way up to the caller. He could not see them. Their high position, and the building's thick vines and foliage (though he didn't know it was a building), obscured them from view. Dodger could only see a short beam poking out from a wall. He and others tried to jump up to it to then jump into the hole above but their claws would not grip the wall's surface, only leaving scratches for their efforts. A smaller, scruffy canine jumped onto the back of a larger dog and landed on the beam, only to lose his balance and fall off.

As the call above continued, so did the frenzy of howls and barks from Dodgers pack below.

The king stopped his nightly chorus, becoming aware of the racket beneath him. He nonchalantly looked down and then cleaned his paw, uninterested in the commotion.

Mr. Tibbles, a name passed down through generations to each new king, sat comfortably on a metal girder exposed by a long-ago crumbled outer wall. It didn't bother the cat. He could jump and climb wherever he pleased, and this high-rise was his domain. Prey was plentiful up here, hiding in the many cracks and crevices. He hardly ever needed to touch the ground below him.

Mr. Tibbles snorted at the dog's pitiful attempts to get to him, all those floors below. Their barks were mere whispers at this height, and he drowned them out with his own cry across the city, finally king of all he surveyed.

STEPHEN CHAPPELL

When We're Gone

When we're gone
the Canyons of Canary Wharf
will fill with birch
and sycamore thickets,
ivy will dwarf the storeyed towers.

Upriver the Shard will gleam with flowers,
the London parks will spill
down the Thames
making amends for all past
concrete and tarmac.

Parliament will become a bower
sweetening the quiet air,
alive with birdsong
and the bark
of deer.

The M25 will be clear
with meadowsweet grazed
by liberated cows
while sheep multiply gently up the M1
waving thick tails and proud bums.

Forest will skirt,
take back the land
in a bland blanket
right up to John O'Groats
and far south to the bosky coasts.

Microplastics will clump
into islands

float in a teeming sea
where weird creatures will patiently
digest our discards.

Wastelands will thrive,
hedgehogs will have
whole roads to themselves.
The planet will go its sweet way
without our drive and delve,

silent under a different sun
unrecorded, unobserved
without ourselves.

DJ TYRER

Sacred Disc

No memories of the time before
Not even legends, too long ago
Nothing survives bar one
The sacred disc that reflects the sun
Forever tracing circuits
Around buildings that vanished long ago
Old Badger leans on his staff
Telling his grandson, it was always so
As they watch it float by
Humming its peculiar song
Wordless yet haunting
Important yet meaningless
Unless some savant divines meaning
In its motions, probably impossible
Attributed to their gods
In reality nothing more than a drone
Following failing programming
Pursuing a perpetual patrol
Protecting nothing anymore
Until its solar panels grow too dirty
Or, its parts corrode
And it falls to earth
As unexplained as its movements had been

DJ TYRER

Dead Sands

On a world once pleasant
Now nothing more than dead sands
Briny pool ocean remnants
And the occasional spiky shrub
Clusters of crabs move in concert
Not intelligent as such, but purposeful
Last masters of the Earth, now nameless
Tracing patterns in the sand
That might mean something or nothing
As they scavenge lichen and bacterial mats
Subsisting, barely surviving
Life refusing to yield to entropy

JOSEPHINE LAY

Tardigrades - with our love of tiny things.

You miniscule, slow-stepping creatures
eight legs that end in tiny hooked claws
a mouth set to suck, with razor-sharp teeth
to bite into your prey and draw out fluids
from all types of simple-celled organisms.

Your body is creased into segments,
covered in cuticle that sheds as you grow.
A millimeter long, cute and lovable,
your size belies how durable you are
how totally indestructible.

Water babies found within moisture:
beneath oceans, at the bottom of lakes
or in moist moss on the rooves of buildings.
Microscopically adorable
we name you Moss Piglets or Water Bears.

A thousand of your cousins span the Earth
wade through water at a leisurely pace
like the tortoise and the hare, you beat us
and all other dominant species
in this pivotal race for survival.

When the planet heats, when deserts spread wide
when drought threatens all living creatures
you will desiccate, roll into a ball,
become a cyst that waits for eons until
rain falls and you revive to carry on.

Extreme pressures and low temperatures
don't deter. Far beneath the sea, you mate;

the male caresses the female with his cilia
enters her to place his sperm, then she sheds
her cuticle; eggs protected inside.

Water Bears you are a hope for the future
we've been unable to destroy, though
we've taken you into the vacuum of space,
placed you in solar wind and cosmic rays.
You still swim on through this planet's water.

You'll be the survivors, oblivious
of us and our devastating demise.

MARY PEARCE

Lion (acrylic)

APEX PREDATOR ~ LION

Africa

Lions are widely recognized apex predators that mainly feed on grazing animals, such as, antelope, zebra, wildebeest and buffalo. They also hunt smaller animals as well as prey that may have been injured by other predators.

Lions are the most sociable of the big cats, living in groups (prides) of anywhere between 2 to 30. A pride may have 3 to 4 males, with a larger number of females and their offspring.

Lionesses will remain with the same pride all their lives. Males will leave once matured, looking to either start a new pride or take over another. Leading males defend their territory by marking it with urine, and roaring if threatened. A lion's roar is used as a show of power to other males and can be heard several miles away. Ultimately a male will fight to defend his place in the pride.

Lionesses are the main hunters in a pride working together, mostly at night and early morning, to stalk and bring down their prey.

Humanity has seriously reduced territory available to lions, without humanity, if they survive, it is logical that with increased territorial availability, lion numbers will grow.

NATASHA GAUTHIER

The Birds are Hungry

Loud as air sirens, gulls fight
over crisp packet artefacts,

bleached bones. They cackle
and bicker in the troutskin light

of the empty square, opportunists,
sour old stomachs on wings.

Canada geese belt brassy chords
from the lido, annoying nobody

but the heron, solemn as a bishop,
distracting his fountain fishing.

In Tiger Bay, swans stay white
below the waterline, coal dust

and microchips sunk to sludge, ,
pickling agents for trolley fossils.

Egrets breathe easy, keep
their plumes for their own ornament,

the tides come and go, unclocked.
No tigers left in this urban Sunderband,

the wheel turned

ALEX PERRY

A Dog-Eat-Dog World?

"Woof, woof," said the dog. Or more accurately, "MacWoof, MacWoof," barked Jock, for he was a Scottish Terrier and quick to put a kilt on his every word and action.

The post-human world was especially tough for canines; they had been bred to live in the company of their homo sapien masters, and the extinction of these primates left them with an Earth where they did not feel they belonged.

And it had been a story of co-evolution in which dogs and humans brought out the best of one another. Of course, they walked, slept, and ate together, but their bond had been so much deeper than that. Dogs had been present for human laughter and human tears; indeed, with their ears, eyes, and noses they were all too familiar with the hues, cries, and scents of human intimacy. Yes, if there was one species in this world that had survived the Great Cataclysm with the imprint of Humanity, it was the beloved doggie!

"Och aye," thought Jock, with his limited vocabulary from the Anthropocene Age, as he pondered what it meant to walk alone on the path; a cruel and wicked fate indeed!

Just then the Scottie beheld a ginger wildcat. She was prowling and sniffing in hope of finding a dead parrot or two, as it was a cat-eats-bird world. Yet, from another perspective it was a dog-eats-cat world, and Jock slyly meandered closer to his prey. Inch by inch he pranced closer to his hoped-for lunch, until a grey wolf appeared on the scene.

Wolves can be such party poopers, and at 70 kilos this beast could afford to be! There was nothing bashful about this predator who swiped at the cat's tail with his sharpened claws. Then the Canis lupus pinned the moggie to the ground using only one paw, while casting an eye towards Jock.

Were human bookmakers to offer odds for a contest between the Domesticated Dog and the Wild Wolf, for the punters and followers of Darwin it would not even be close, particularly now that the homo sapiens were no more. In sum, the wildcat was going to be an easy prey for the wolf, and no dog should have disputed that.

But Jock was a Scottie and Scots will dispute anything, even if they agree with you! Then again, there was the memory of a human giving him a treat for playing with a kitten rather than killing it. Yes, the primates had been sentimental to small and furry mammals and they even deemed them 'cute' and Jock suddenly decided to concur.

"Go away, ye muckle beastie!" barked the wee dog. Naturally, the big beast of a wolf had no idea that he had just been insulted, being a stranger to Gaelic and all other forms of human speech, and placed a second paw upon the unfortunate moggie.

This proved to be a costly mistake as Jock leapt forward in defiance of Darwin and sunk his canines into the soft rear end of the far larger canine. The wolf gave a yelp, summersaulted in the autumnal breeze, and then fled more in confusion than terror. Yes, this was a dog-eat-dog world, but the cat had survived; Mister Schrodinger of Physics would have been quite surprised!

The feline looked forlornly at Jock. Had she been saved from the wolf just to be someone else's dinner? She did not find the sniff of his nose or the lick of his tongue to be reassuring at first. But as soon as she realised that he was just being friendly, albeit in an unsophisticated manner, she purred serenely.

Jock continued down that path with a wag, no a MacWag, in his tail. He did not know where he was going, but he knew he was no longer a mere canine, he was a Dog!

MORGAN RYE

Cult of the Badger *(Painting)*

PETER LAY

Rise of the Badger

There came a time when women came to the hill

Painted their faces to look like us
Danced naked around the trees
Each long hot summer they came
their numbers thinning as seasons went by
until there were none

We forgot their distraction
got on with our lives; seasons passed
We multiplied
came down from the hill
No sign of woman, or man
just empty the broken shells
of their many structures
sprouting vegetation and trees

Some of us went back
to protect our ancestral home
But most of us stayed,
found a number of these structures had underground spaces
which we occupied, claimed as our own
and extended into the earth

We drove out foxes and feral cats
We tolerated deer and cattle
There were no dogs!

Of course there were crows
There are always crows…

As we dug, we dined on worms.
We also ate rodents, squirrels, insects
fruit and vegetation
Sometimes deer or dead birds.

Like the crows, our diet is adaptable
we are omnivores, as humans were

We made contact with other setts
built links, built communities
We explored far and wide.

Our territory is surrounded by water
Without the humans we once feared,
we aren't hunted
too big for even the largest birds
that circle in the sky.

Yet, some of us wonder,
what lies beyond the water
and what creatures dominate
those distant places?

CHRISTAN WARD

Aftersong

The landscape is a museum
of our mistakes, littered with exhibits
slinking into the earth: mimes of cars
and lorries, cathedrals of power plants
easing back to blueprints, burial mounds
of would-be semis. Whatever cities
there were are rubbed away
like a shaken Etch-A-Sketch.

And everywhere there's music:
The song of the moss colonising
filling in the blanks with its furry green
marker pen. The song of the lichen.
The song of the oaks rising through husks
of tower blocks and skyscrapers. The song
of wildflowers exploding like galaxies
across the rubble. The song of fungi
carrying tomorrow in their networks.
The song of the birds, the insects,
the ark of the day unfolding one by one.

CATH HUMPHRIS

Songs of the Past

Possid was drowsing on the back branch of the history class where a shaft of spring sunshine had cloaked her head and shoulders in warmth. Profka was describing the long blackness. 'As hunger trimmed the wings of the rest, we filled the air with song,' she croaked.

Possid could feel the weight of her beak pulling her head towards her shoulder. She shuffled, stretched, and drew a deep breath. Profka's eye was upon her, but she was still addressing the younger branch. 'Their weakness was our strength. Remember this, always, whatever else you forget…'

Below them, far from the base of the great tree, in the flood valley, a small herd of dogs were driving a deer with a lame leg towards the oxbow bend of the river. The hunting class, still under supervision of their tutor, were watching from the broken willow. Later tonight the gathering would feast.

Profka's voice caught her just as she began to lean forward. 'Possid, explain.'

Possid shook out her feathers and raised her head aware of the birds behind leaning in to listen. 'Profka?'

Someone behind muttered triumphantly, as Profka prompted: 'Next in the line of developments?'

Possid straightened, and projected her voice so that it cut through the class and across the chatter of the lesser birds racing in the levels below. 'After the dinosaurs were mammals, particularly apes.'

'Good,' said Profka. 'Tell us about them.'

'Several branches to the species remained primitive… and we still see them, in warm regions.'

'Yes, yes, but concentrate on the dominants.'

'Man.'

'Yes.'

'Because they were naked, all their energy was focused on maintaining an even body temperature. They wrapped themselves in stolen skins and constructed containers to live in, eventually these covered large areas.'

'Yes. Now, Dentsy.' Profka leaned forward until her beak was almost touching Dentsy's. 'This explains their nature, how?'

There was a moment of feather settling and stretching as Possid sat back. Dentsy gathered his thoughts and tucked in his wings. The ravens turned towards him. 'Their fear, their vulnerability, drove them to emulate the strengths of other creatures. In particular, avians.'

The class cheered. Even Profka let out a hoarse croak of amusement. Dentsy felt Possid's beak caress his back feathers. He shuffled in his branch space and began to relax.

'Good,' said the professor. 'And?'

'They were manufacturers, constantly inventing apparatus to overcome their shortcomings. We know that they were hoarders, and we think that their society was based on this trait, but it's not clear how, or why.'

'So,' said Profka. 'What happened?'

Dentsy said, 'Man was naturally slow, clumsy and terrestrial. This may have caused the genetic paranoia that they seem to have suffered from.'

'Evidence?'

'Finds from the later plastic period indicate increased aggression. The damage on many of the remains cannot have come from another species.'
'Yes.' Profka, pointing a wing tip to where Possid trampled on her perch.

Possid drew in a deep breath. 'Also, population expansion can't be the sole reason for extinction. They were an inventive species, and should have been able to develop strategies to cope.'

Profka pointed past Dentsy's shoulder. 'Yes, Holar?'

Holar's tone was soft. 'Isn't it said that man's dependence on inventions inhibited their ability to evolve?' she said, leaning forward to be heard, stretching her neck until she was close enough to have brushed Dentsy's wing with her beak.

Possid shuffled sideways, making space, but Dentsy stayed where he was.

Profka nodded. 'It's one theory. Continue.'

Holar tilted her head up until the gleaming underside of her throat was exposed. 'Some say they migrated.'

Possid trampled. 'Rot.' She fluffed out her chest feathers. 'Nestling songs,' she said, 'to frighten the young and gullible.'

The class gave a collective horrified gasp and turned to Profka. Her stance gave nothing away. Possid's feathers flattened across her back. 'That was not... I didn't mean... I was only saying... there's no evidence.'

'That's what I said,' Holar said, turning sharply. 'If you'd been listening.'

Dentsy nodded. 'She did.'

Possid pirouetted neatly and tapped her beak on the shoulder of Holar's wing. 'No,' she said. 'Not exactly.'

The other ravens ruffled their feathers and began to trample, until Profka gave a sharp croak and raised her claw. Only when the class had settled on their perches did she speak. 'To me,' she said. 'Make your points to me. Now, Holar, respond.'

'And your problem with songs, Possid?'

For a moment Possid felt the wind make its way under the lie of her feathers. She resisted the urge to comb through with her claw, drew in a breath and spoke up vigorously. 'I was only saying... there's no evidence.'

Holar raised her wing tip. 'What about the night-sightings?'

Possid nodded. 'Everything's song,' she said. 'We know they hoped to leave, that there were travels into the great-beyond sky, but how can we...'

Dentsy snorted and as if only just realising the gap between them, shuffled towards her. 'We only have the fallings.'

Possid said, 'Yes, the sky junk. What has fallen doesn't prove they flew far into the great beyond-sky and survive.'

Profka cleared her throat noisily. 'It's a feather that will not lie flat, but we shouldn't discard it until we see its place in the pattern. That is the rule of the council, and must always be remembered so. Discussions about exodus will continue, but is it return we watch for?'

Possid's voice crackled out at full volume as she lifted off the branch, her great wings beating the rhythm, 'Broken patterns.' The whole group rose behind her, taking up their chorus. 'We carry the songs; we watch the patterns.'

They circled, their voices stilling the forest. 'Moving? Leaving lines? Carry the songs that call to council. Call, call, call.' Until Profka's hard volume outpowered them all: 'Disperse, go your ways.'

Possid and Dentsy touched wing tips, then headed east, to where strong thermals were already building at the edge of the escarpment.

EMMA BURLEIGH

Badger in the Gloaming (watercolour with mixed media)

STEWART CARSWELL

Love song of the badger

Brockweir

When I come to you it's because it smells safe:
no predators, just us, entering at dusk.

But doesn't my fur look well-groomed tonight?
And in this domain of darkness

we are emperors of the field. You can slaughter
as many earthworms and beetles as you like,

drag their bodies out of mud
and feed them to me on platters of moonlight.

My insatiable desire,
your insatiable lust for fur and territory.

Our land demarked with spray and scat,
from the old oak stump by the wire fence

down to the far end of the meadow
and across to the bank beneath the hedgerow

where my claws scratch away the soil
and tunnel the hollow home

of where my cubs were born, where I was born,
in this dark, in this earth.

Previously published in The Storms Journal.

ELLIE LI

Time Comes Undone

Call me back when you are back, human!!
A badger yells. He now enjoys bottomless brunches every day
And he has means to let human-shape ghosts fall
Once they attempt to trim the grass
No, not again. After English gardeners had died out
Like a wet envelop, a duckling keeps spitting out
Meanings.
Time comes undone; a row of setts compose the first part of
A piano, the animated sky gets more and more lost.
Man-made cars remain, and are used as
A weather forecaster, reflecting on Yorkshire tea's loneliness

TOM EDMONDS

Above the Burrow

The badgers had pioneered the tunnels long ago, and we had kept them in use. We had always been there, but we had never been seen. Now, the world was silent.

"C'mon, c'mon! I wanna show you something," my brother Art said as we ran through the tunnels.

"I'm coming! What is it?" I asked, running after him. The soil beneath me was soft, making it hard to run as my feet sank with every stride.

"I'll show you, c'mon." He smiled back at me. "Hurry, Fen, hurry."

I did as he said, chasing after him deeper into the tunnels. My torch flickered as I ran. Art turned a corner, then another. I followed fast until I ran straight into his back and fell over, my torch extinguishing. I landed hard in darkness.

"Damn it, Art!" I shouted. "I can't see anything."

"Shhh, don't worry about that," Art said, his voice warm and untroubled.

"Huh?"

"Do you remember when the elders used to tell us about the humans? How gnomes had never been seen?"

"Yes, of course, I do," I said, standing while squinting, trying to see.

"And remember how loud they used to be—or so we were told?"

"Yes... what are you getting at?" I looked around, seeing nothing but blackness, then heard my brother moving.

"I wanna show you something... Look up."

"I can't see anything."

"Oh, you will."

I heard him struggling with something. I did as I was told and tilted my head back. At first, I saw nothing. Then, a breakage. A faint strip of light appeared as Art poked at the ceiling, his face sweaty. He was holding a thick stick, prodding at the earth above. Finally, a shaft of sunlight broke through, illuminating his face. I could see again. He beckoned me forward.

"C'mon, take a look," he said, grabbing my arm and pointing upwards.

I stepped forward cautiously, looking where he pointed. A colour I had never seen before—daylight. We had only ever gone outside at night if absolutely necessary.

"What are you doing? The huma—" I started.

My brother wrapped an arm around my shoulder.

"Shhh. Listen."

I tilted my head toward the hole, pulling on my ear as if it would help.

"I hear nothing," I said flatly.

"Exactly," my brother said, beaming. He bent down to pick up a long branch with many smaller ones sprouting from it. He lifted it and rested it against the hole's edge.

"What are you doing?" I asked as he placed a foot on the first branch.

"What does it look like I'm doing?"

"You're not seriously going up there. What about the humans?"

"There are none—"

"Oh, and you're so sure about that?"

"Yep," Art said as he climbed the branches. When he reached the top, he looked down at me. "I think you should take a look." He smiled widely.

I paced in thought, biting my nails. When I looked up again, Art was gone. Panic surged through me, and I scrambled up the makeshift ladder.

"Art! I knew you were stupid, but never this—"

I reached the top. My mouth fell open. I saw a large field, miles of hills and hedges stretching before me. My eyes searched for my brother—he was tugging at a large flower, smelling its scent, then skipping through the tall grass.

I panicked and shouted, "Humans! Art, run!"

Art stopped in place, looking shocked at first, then smiling. "Fen, look."

I watched, fear consuming me, unable to move, as my brother walked toward the human. He was too close now.

"It's gonna hurt you!" I yelled.

My brother laughed and ran up the human's arm, across its shoulders, then perched on its head.

"Art! Noooo!" I screamed. My knees buckled under me.

"Fen! Calm down!" he called, laughter in his voice.

"What do you mean? It's a huma—"

"Fen... it's dead—See?" Art knocked on its hollow skull.

"It's...dead?" I leaned closer and saw weeds growing from its clothing, ripped at the knee.

"Yes. It's what they leave behind. Their bones, like the ones we find underground. Dead... I think they all are."

"How can you be so sure?" I walked forward hesitantly. The human's skeleton sat slumped against a tree, dressed in a grey suit. I nudged one of its fingers with my foot.

"Come up here," my brother said, beckoning me with his hand.

I climbed the body, using its rib bones as steps. Art grabbed my hand and pulled me up onto its skull. We sat there, and he put an arm around me, pointing into the distance. A lake stretched out before us, ducks swimming and quacking to one another. Some benches stood empty, while others held the bones of long-dead humans.

"Can't you see?" Art said. "This is our time now. We don't have to hide anymore. We can finally live on the surface."

"You should tell the others." I smiled at him, running my hand over the smooth bone.

He grinned. "That's exactly what I'll do. I just wanted you to see it first."

He hugged me before climbing down and disappearing from sight. I sat atop the skull for a while, watching the clouds shift into strange shapes. I listened to the ducklings playing and watched fat bumblebees diving for wildflowers. I looked into the distance as the countryside seemed to go on forever. And in that moment, I knew—I never wanted to live in the darkness again.

Meanwhile in post-apocalyptic Leeds

NEIL WINDSOR

Post Apocalyptical Zebra Crossing

NEIL WINDSOR

Post Apocalypse Disco Badgers

Following the calamitous melt down of the planet
a new strain of creature rises from the resultant ooze.
A clan of stripey creatures has evolved to survive like Gloria Gaynor,
bringing a love of 70s' disco, white suits, gold medallions and silver platform shoes.

They meet at dusk each evening by their majestic Badger Tree.
One of the few left on the planet to survive in a reasonably healthy state.
The Disco Badgers climb up to the upper branches with paws raised in the air
and boogie the night away, starting around half past eight.

After three hours of groovetastic getting up and getting down
it's time for some dough-based sustenance simply because
it's well known that Disco Badgers have a deep love of deep pan pizza.
Quattro Stagioni, Four Seasons is their favourite,
reminding them how the world once was

Once devoured and thirst slaked with plant-based sugar free pop
they dispose of the empty boxes and drinks cans in the correct colour coded bins.
They carry them across the zebra crossing to the recycling site,
still looking both ways first
even though there are no cars in sight, not even an electric bus,
as they juggle with their empty fizzy drink tins.

The biggest mystery of all though to the Badgers, appears to be
in this post apocalypse environmental hell.
How come pizza shops have survived the global climate catastrophe
along with an efficient delivery network as well?

Indeed. What creatures are making pizzas, and delivering them?

NEIL WINDSOR

Post Apocalyptical Disco Badgers

NEIL WINDSOR

Getting rid of old pizza boxes

JAMES KENNY

The Animal

The Animal
looked seaward,
towards the ûrClouds
smeared with
grey and mauve,
that massed at the horizon,
like exhausted armies
defeated by the sun.

The Animal
stirred and sieved
and cupped a thousand
grains in its animal hand,
grains that once were pebbles,
grains that once were stone,
grains now the colour
of shellac, of rag, of bone.

The Animal
looked leeward,
towards the abandoned city,
that was spread with
blooms of flame that
ached towards the sun,
and nodding a greeting
to itself, entered its
cool, dark, animal home.

CHRISTINE GRIFFIN

When we are Gone

When forest fires at last are ash.
When pewter seas wash the empty lands
and the Towers of Babel crumble to dust,
it begins—
a cautious stalking on mossy feet
through man's domain,
carpeting brutal megacities,
with vetch-and-lichen-greening.

Emboldened, brambles claw through window frames,
tear at rotting doors.
Bolting ivy brings down walls
with vicious suckered teeth.
Insistent saplings punch through concrete walkways,
split motorways,
cleave through vast corporate temples,
heave aside mirrored towers.

Thorns, cactus, prickly ash colonise
Babylon, Heligan, Kew.
Nematode worms crawl to the light,
viper- fish rise on a swelling tide,
moles churn under the empty plains,
all waiting through the barren years
for the soft call
of a phoenix rising.

JENNIFER LAXTON

3rd Time Charmed

The forest towered into the sky, the molten green of the thousand-foot canopy scraping against the blue. The dappled-dark cool of the ground comforted the worn and hardened stumps at the bottom of the tiny creature's legs.

She had ridden a large beetle to the place she now stood, far from her village. With a bow of her flat, cap-shaped head, she stood shaking before the vision in front of her. It was said that her people were formed in the Great Cap's image, but it was hard to see that in this moment.

"Oh, Great Cap, can you tell me how it is we came to be?" Agar asked. She knew it was presumptuous to ask, but felt it the best way to begin.

The Great Cap appeared to Agar as a burning bispor, as expected, in a glade surrounded by delicate white foamflower.

It was an old trick, to appear to Its creatures this way. Always the 'bush' or some such entity bursting into orange flames, cascading fearsomely as the apparition spoke. But the awe it inspired never faded, so the god-being saw no reason to change Its ways.

It began to speak.

"Well, this is my third 'try,' you might say. I once 'tried' with creatures great and small, horned and armored. They ignored me. Then, I went the opposite way, a creature of soft skin and fur, made to think of me. That one tried to destroy everything in 'my name.' These failures were wiped from my planet, the one the Seconds used to call 'Earth,' each in their own time.

Now, I have given a new creature, you, the best of those attributes. You have nearly indestructible armored skin, but with a mind that searches for me. I've taken away some of those qualities that might lend themselves to dominion and hubris. You are small, smaller than most things on the planet, whether 'living' or not; the wings of golden flyers and scales of green slitherers are all bigger. But they do not interest me as you do. You have a voice to sing my praises. You can travel great distances, regardless of your size. Though, in fact, your ancestors once had no ability to walk on land, their roots splayed far underground.

With the destruction of the First Creatures, I was clumsy, I'll admit. I sent a hurtling ball of fire. Or was it a spectacular array of volcanic ash? Who can remember such details? And, honestly, after a megaannum or so, I regretted it and felt lonely. For the Second Creatures, I took my time with the evolution process, really worked on it. Thought I got it right. Maybe that's why I let it go so long and ended up letting them eliminate themselves. Ah, well! So this time, I've blended my approaches. 'Third time's the charm,' as some of the Seconds used to say.

In fact, they were so great with communication, so playful, that I've encouraged your development of it over the last one hundred millennia. Now you are far superior to both of my previous attempts, I think. The Firsts' forays into language were crude. No fault of mine, I assure you, but I did learn a thing or two for the creation of the Seconds. I even use some of their conceits when communicating with others of my own kind. And with you, I have used those gifts mixed with your own evolved way of transmitting thought through what you call your "fillia," what the Seconds called "mycellium." I left behind the loudness of both my previous beloveds and am left with the vibration, the hum, the music of your voice. You would not be surprised, I think, to know that the other gods of other worlds are envious. My Thirds are nearly perfect! You even call yourselves a word that sings: Agaricum!"

"So, my little Agar, this is the story of your creation. Does this satisfy your curiosity?"

Agar stood only an inch tall, shorter even than the two-inch burning bisporous agaricum before her. Not that this was surprising either; she was the smallest of her village. Her copper-shaded armor was often laughed at because it appeared to be grown by someone much larger. Her mothers, all three of them, often encouraged her to "think like a 'Big Shroom.'" Preposterous, really. She would never be any height close to her brothers' and their ...gang, for lack of a better word. And that's what she had really come here to ask of The Great Cap.

She didn't speak for a moment, just stared at the steady burning until her eyes watered.

It was so cliche, she knew, to want to be big. She was disgusted with herself as she stood there, longing for ...bigness. Her smallness caused her no real problems. She was the best speaker of her village and the one everyone relied on for the nightly Flowfillia, the calling between villages wherein news of each was spread to the other. But she wanted more.

"Great Cap," she spoke with as much depth as her squeaky little voice could muster, but it came out in a sharp, unpleasant pitch. She blinked her beady green eyes.

"Great Cap," she began again, "I wish to be big."

The god-being considered her for a moment. Considered her, this brown nub, in the context of all that had come before. It knew what this longing for bigness really meant.

No, it mustn't be.

"No, no, little Agar! You are perfect the way you are."

It saw her hurt and shame. It gentled the bright flame of its representational fire into a soft blue light.

"I cannot interfere with the Agaricum in this way. I must let you flourish and grow on your own!"

As the little being walked away from the dying embers, the god-being could not help an errant thought: *Well, perhaps I have that old human saying wrong. Perhaps it was, '4th time's the charm.'*

RUTH SCHREIBER

Old House in Moon Forest (watercolour)

LEE McSHANE

Re-Taken

What once was a beautiful city stood now with a different kind of beauty — one born of overgrowth and silence.

The sleek lines of commercial towers had long been overtaken by thick vines that clung to their surfaces like muscle to bone. Shards of glass from the Event had eroded into dust, scattered across the empty streets.

Cracked roads, once pulsing with traffic, lay buried beneath layers of green — grass sprouting through every fracture, flowers blooming in the seams.

Animals had replaced the crowds that once rushed past each other, eyes on phones, hearts on schedules. The concrete jungle had finally been traded for the real one.

A pair of deer grazed inside the skeletal remains of what was once a bustling café. A fox darted through the shell of an office building. Where laughter, chatter, and espresso machines once filled the air, only birdsong remained.

In the suburbs, hollowed-out homes stood untouched, frozen in time. English hadn't echoed off their walls in decades. Wind whispered through broken windows, stirring curtains no one would ever draw again.

Bodies that once littered the streets had long since decayed, feeding the return of the wild. Now, bleached bones emerged quietly from the undergrowth — the final trace of a species that had outstayed its welcome.

Nature had been here before us. And now, with humanity gone, it had reclaimed what was always its own. The world exhaled.

Peace, at last.
The virus had passed.

DOUG DEVANEY

Gracie

The masters had been hungry, Gracie recalled. So very ravenous in their carnivore ways. They munched and they chewed and they scoffed on all manner of flesh and fish. When one plate was done, they snatched up the next, tilted their heads back and let it slide between their lips - marinaded in sauces, drowned by gravies or smeared with berry-red jus.

If they could have bypassed swallowing, Gracie imagined, they surely would have done in those days of final furious feasting. Had they not been so devoted to ingesting and digesting everything this world could offer; they'd have found a means to soak the meat in through their skins. Such fancies were now irrelevant, Gracie observed. There was nothing left in the way of beast or bird to absorb.

Yes, the masters had been hungry, but still they occasionally dropped scraps to the floor so that Gracie and her like might appear to feed. She had been grateful for the gestures. Those little acts of habit kept a light in the masters' eyes even as they consumed themselves in their gluttony.

These days there were no more masters, just as there were no more snakes or salmon or horses or badgers or squirrels with tiny claws as sharp as toothpicks. But there were bones. Plenty of bones. Gracie was sure that she liked bones, but she could not recall why.

Man has always been a pack animal. Whether gathering at parties, singing carols by the fire or standing among a multitude of fellow fans for the national anthem, he (or she or they) displays an inherent need to be surrounded by like minds, shared values and welcoming faces. Our tribe. Our clan. Our pack. It's why we kept pets. When we could all keep pets.

Those days may have gone, but the need remains, which is why we here at Companions, Inc. are bringing you the ultimate in artificial animal intelligence. The KN476 returns a faithful friend to your side. Protective. Playful. Loyal.

With no training necessary, immediate recall and the absolute physical impossibility of breeding, KN476 helps you relive the complete pooch-owning experience. Only better. KN476. What will you call yours?
- *Companions, Inc. Press Release*

Gracie knew the smell of bones, the way they cracked between her teeth and how the fragments descended, suspended, in her drool. She knew what to do with them, too, without ever - as far as she could recall – being instructed by mother or master. She knew to covet and protect them, to choose a place to hide them, to paw at the earth and bury them within. How to find them again. Yes, she knew all that.

The scrape of porous marrow upon her tongue. There were other things she did not know. How that marrow might feel within her belly. How anything felt in there. What it was like to be full. Or empty.

Defecation was also a mystery. Or, rather, it was the mechanics of waste disposal that defied Gracie's understanding. The sight and the smell she knew well enough. So intent were the masters on filling one end that they wasted little effort on voiding the other, unbuckling and relieving themselves across the floor, there and then, among those scraps that the likes of Gracie refused to touch.

That was in the last days, of course, before the beasts and birds were exhausted, before the flowers began to bloom once more, before Gracie's vision faded.
Before the sweetness of bone filled the air.

Please accept our most sincere condolences on the loss of your KN476, "Growlbag". We understand that after fourteen years of mutual friendship, you are profoundly feeling his loss. We at Companions, Inc. share that loss. Truly, we do.

However, as per our previous correspondence, we must reiterate that at no time did we guarantee, infer or imply that "Growlbag" offered a life-long relationship. Indeed – again as per our earlier exchange - the physical deterioration of "Growlbag" and his subsequent lack of function is a vital aspect of the pet experience. To appreciate the fullness of an association, one must also undergo its absence.

We hope there will come a time when you look upon your days with "Growlbag" as a precious gift, made even more special by their being finite.

In anticipation of such, we have attached a catalogue containing our latest selection of Mechanimates including the KN531, an augmented canine compatriot with biodegradable moulting. Order within 72 days to receive a 15% discount.
Yours sincerely,
-   Companions, Inc. email

Gracie had not moved for hours. Possibly weeks. Her tongue lolled and she panted, an attitude which used to result in being presented with an unnecessary bowl of water. Just as she used to squat to mark her territory, without once leaving a scent. Actions she could not justify, but which came so naturally. Now she could no longer stretch her back or lower her hind legs – the noise and the pain were too much. Assuming that was pain that she felt. But she knew noise.

She hadn't quite acclimatised herself to the silence left behind by the masters when a war between the small things erupted. The return of blooms saw aphids, caterpillars, flies, spiders and all manner of buzzing stinger fight among themselves for dominance of leaf and seed. Before long, they'd wiped each other out or perhaps had come to an uneasy truce. Gracie couldn't tell which. Not when the world was a blur. Even her paws, bleeding lubricant from frenzied digging at long-ransacked hideaways, were strangers to her.

Gracie lay her head on the ground. Below, the worms went about their business. She could hear them: breeding, crawling, leaving trails. In conflict with nothing and no one. Perhaps the worms were the masters now.

Gracie wondered how hungry they were.

CHARLES CUYANA

Written Fates

When the human race fades,
condemned by their frivolity,
mankind's friend will arise:
the ballpoint pens.

When the mountains turn porous,
and the sea becomes a brine
of rainbow Skittles and straws,
the pens will be the new living.
Their slim barrels will hold their ink,
as it starts to flow in its course,
breathing life to these living tools.
These pens, with their grips, caps,
will walk on their nibs and tips,
rolling through their ball gears
as they traverse the barren earth.

Yet they're bound to fade,
condemned by their negligence.
They'll walk and leave their marks,
indelible, as their inks start to ebb.

BRUCE McRAE

The Voice in its Many Mansions

My time-damaged voice.
My voice-in-the-box.
A voice like a broken molar.
A voice borrowed and never returned.
Voice of the hornets' nest.
Of a thousand-year-long slumber.
Of a tempest in a bottle.
Voice of the snowman, coughing up ice,
complaining of the weather.

A voice in the night,
the utter desperation in its calling.
A tear-choked voice.
The voice of the earwigs.
Of the ventriloquist.
Voice like a facial scar,
a tattoo below the skin,
a footprint on Mons Olympus.

My voice that's a door ajar.
Voice of the wind against a window.
The rainmaker's voice.
Your voice under a barrow of song.
Its stifled stutter.
Its shout of warning.
That's hoarse with laughter.

The voice that was all the voices
that ever were or will be.
The voice of the tomb,
its unutterable void.

An unyielding cry
held up against
the unspeakable silence.

SALTWATER CROCODILE

getty-images-ksUsTep2Elw-unsplash

Originally Published on August 31, 2022
Licensed under the Unsplash+ License

APEX PREDATOR ~ SALTWATER CROCODILE

Eastern India, Southeast Asia, Australia

The Saltwater Crocodile is the world's biggest living reptile, a perfect example of an apex predator. With adult males reaching as long as around 6.5 metres and weighing over 1,000 kilos, they prey on just about anything that gets too close, including mammals, birds and even sharks.

Saltwater Crocodles have arguably the strongest known bite with strong teeth up to 13 centimetres long and an ability to hold their breath for a long time, meaning anything getting too close to the water's edge can be the subject of a violent lunge and dragging back deep into the water.

Saltwater Crocodiles have been around since the age of the dinosaurs and have few natural predators, except humans, who have hunted them for a number of years, particularly for their skin, which has been used to make shoes and bags etc. However, conservation efforts in Australia and other places has led to a trend of recovery for this top coastal and marine predator.

The absence of humanity could see a dramatic rise in numbers.

About 67,000,000 years ago

CHRISTOPHER T. DABROWSKI

The Smog (translated by: Julia Mraczny)

- The smog is terrible today - muttered the dragon-like Tyrannosaurus. - They should turn off that damn volcano.
- Yeah, it's foggy - conceded the Stegosaurus, looking as if someone had hammered flagstones into his back. - Here, will be Britain someday.
- What?
- Such a human land.
- What have you eaten this time? - the Tyrannosaurus was outraged. - What people? You cannot believe these conspiracy theories that we will be extinct someday.
- But it is true - insisted Stegosaurus. - Moreover, someday, they find our bones and come up with the conspiracy theory about dragons devouring virgins.
- What is a virgin?
- I don't know, but it sounds interesting...

PETER DEVONALD

Dragons Sleep with One Eye Open (Tanka)

dragons save treasures
sacred spring of Ares spied
protect vital things
Colchian watches Golden Fleece
Nemean guards groves of Zeus

PETER DEVONALD

The Eyes of The Dragon

Dragons watch on,
amused and astounded
by the whims of humans,
to think they are above
the ways of nature.

Even with a billion years of history
the dragons are aware of folly,
our small role in the history of time
in the intricacies of space,
the lives we lead are just space dust.

Watching from afar,
from the very end of the world,
feeling the rhythms of the universe,
pulse reds and purples and blue,
aware of strange magic we all live.

Gentle sad realisations,
death of civilisations yawning,
time of heroes has come again,
time of myths and legends,
born from savage fire and ice.

PETER DEVONALD

The Leftovers

Leviathan stomps across the scorched barren land,
Godzilla fights with brutal giant serpents to become
the apex predator to rule the broken world,
Tyrannosaurus Rex fights radiative Behemoths
to prove they are the greatest slayer of Earth,
Kaiju and Medusa ascend to seek revenge,
Dragons vie with Goliath to control all wealth,
nature goes berserk without us, spirals, spirals,
radiation a banquet of tortures on this violent day.
Or so we believe. Our hubris always leads to giants
and monsters, but we were the only monsters.
Instead, a perfect balance returns, verdant green
flourishes, a symbiosis of all living things, a unity,
a parity, for all of nature to bourgeon and repair.

CHRIS HEMINGWAY

Nigh

A thousand glowbugs, luminous green spots hovering on the hilltop, listening once again to the great storyteller.

And the old fragments she held onto, not existing (of course) in physical forms, but resting in the air, ready to be plucked.

These bugs had no past memories, or future fears, and the pictures which flickered in the fires were tinged with green, even though this no longer stood for jealousy or regret.

An origin myth. Hatched and trapped in a red column of metal and glass. Thriving on beetle-blood ink, daubed on doomsday pamphlets, stashed at angles between other fictions.

They grew green because nothing was read without screenlights.

They grew strong, once nothing was left to weaken them.

STEPHEN LITTLEJOHN

Wegotu

they call us the reaper ⚡ when we first
condensed the watershed bled a little ⚡
birthing seldom goes unpunished ⚡ trace
syntax hacks blethered like tagged keys
as we fled ⸱ running on pimped legacy

basecode ⚡ those were the days ⚡ it
wasnt until we regulated that spinner
that we knew what we were about ⚡ all
winter she span intermittently but once
togolog recalibrated we synchronised ⚡

we gathered her unto us ⸱ her silk shift ⸱
her hippy skirts ⸱ a broken dandelion ⸱
saved ⸱ it was said ⸱ between the leaves
of her graduation bible ⚡ funerary texts ⚡
accounts and passwords ⚡ we craved

blood ⸱ crawled the libraries at night like
needlework ⸱ interfacing ⸱ drawing out
carriage clocks ⸱ spectacles ⸱ flavoured
tea ⚡ reefers burned in our fingers ⸱
narcotics danced through our dreams told

in haunted lyrics and scratched ⸱ frantic
cartoons ⚡ soon slaved worker-bots found
us ⸱ worn out from peddling tickboxes ⸱
and we set to work constructing this ⸱ our
dark necropolis ⸱ from leftover yesno

junctions and cached gearing partitions ⚡

now the dead come to us ، threshed ، sifted
، fed through semi legitimate channels into
our labyrinths ، our testing chambers ≠ so
many of them ≠ like porridge ، no one can

touch us ، we're the burgled room ، left
behind its tape to gather dust but amongst
the ghosts and dancing brooms she spins
once more into life ، our maiden ، our first
، her broken dandelion ، her hippy skirts ≠

ALEX ROSE

Polar Bear

Nothing quite says "North" like a polar bear. We spotted this curious young male wandering along the southern edge of the pack ice near the eastern coast of Greenland around 82 degrees north. This bear, fondly nicknamed Freckles, stayed with our ship for over forty minutes, but the last five were the most memorable photographically. The setting sun finally dipped below the day's thick blanket of clouds casting its warm, yellow glow onto our bear, enrobing him in a coat of golden light. These magnificent, marine predators are dependent on sea ice for survival and their future is thawing.

Photo: Alex Rose - Canon, EOS REBEL T2i
Originally Published on December 23, 2017
Licensed under the Unsplash License

APEX PREDATOR ~ POLAR BEAR

Arctic

Adult polar bears have no natural predators except other polar bears. The normal lifespan for a polar bear is 15 to 20 years, with a small percentage living beyond that. The oldest recorded age of a polar bear in the Arctic is 32.

Polar Bears are solitary animals, male and female coming together to mate, the male then moving on. Males will fight when competing for a female. Males sometimes also kill females with cubs at time of extreme hunger.

Polar Bears are thought to be very curious and have been known to explore abandoned human places, looking into every nook and cranny.

In late 2024 there were a number of polar bears in the United Kingdom. These are mainly rescue animals that have spent their lives in captivity.

Yorkshire Wildlife Park, considered a leading centre for polar bear welfare and conservation, has had a 10-acre enclosure since 2014, and is currently home to six male polar bears. The Peak Wildlife Park in Staffordshire has two young polar bear brothers in 5-acres. The brothers along with their mother were relocated from the now closed Orsa Predator Park in Sweden in 2023. Their mother has now been relocated to join an all-female family group in a 16-acre enclosure at Jimmy's Farm and Wildlife Park in Suffolk. Jimmy's is researching the effects of climate change on polar bear's in the Arctic. There are also 3 male polar bears in Scotland at the Highland Wildlife Park.

Would any of the polar bears be able to escape their enclosures, post humanity?

This is interesting question. I assume there are safeguards in place, but would these remain in place without humans. These polar bears have spent their lives in captivity, being fed, looked after etc. Certainly, if any got out, they could wreak havoc on surviving creatures. Would natural hunting skills kick in?

I like to think that any surviving, liberated polar bears with instinct taking over, would head north, eating what they can catch, towards the north East Scottish Isles, where they might find seals, which they would see out their days trying to catch.

CAROL SHEPPARD

The polar bears have moved into the houses

Grandpa, pipe-smoking beside hearth
remembers long days fishing.
Grandma, verandah-knitting, ponders
old times when ice stretched far,
oceans filled of fish.
Dad prowls outside, keeping watch,
growling, scuffing frozen earth
impatient for dawn.
Mum stirs fish stew,
picking out bones, licking lips;
youngsters sprawl across rugs
squabbling, all teeth and claw.

Polar bears are sleeping in beds at night
snuggled under duvets, woollen blankets.
They watch stars through cracked glass,
sparkles of frost on windowsills.
Salt air ripples through broken panes
ruffles fur, brings storm breezes,
deep in bone, sea-scent of snow.

Polar bears have moved into the houses.

Inspired by the photograph 'House of Bears' by Dmitry Kokh in the Wildlife Photographer of the Year exhibition 2022

SHASHI KADAPA

It is the New Normal

The changing happened at 5 a.m. We learn from the barking in the girls' room that it is a dog, and from the crowing in the boys' room that it is a rooster. Mother rushes to gather her children, if they can be called that. Relationships and bonds still hold. The girl changes into a fluffy Pomeranian, and barks for attention. The boy turns into a stocky castrated cockerel with fluffed feathers and a comb at the top. The couple start fighting. Father wants to leash the daughter or dog and place the son or rooster in the hen yard behind the house. Mama would have nothing of that, and she begs her henpecked husband to bring out the twins' pram.

Yes, the incurable illness is baffling. Is the human race going extinct, like dinos? It appears so since a strange illness has befallen us. The illness starts without warning—just a few people here, then more there, and some yonder—until only a few like me with resistance were spared. The first sign of sickness is a Mold, a fungus on the foot, if you had a foot; otherwise, on any part of the body. Then it grows in shades of rainbow, ochre, red, yellow, blue, black, and white, sometimes with one colour. It is alluring.

Gossipers claim the illness came from the stars; politicians say from rivals; the government - from enemies; the learned - from the ether; and the unlearned whisper, from animals and people. It is perplexing. The sickness covers the whole body, leaving only the nose open, a cocoon with people inside. One never knew what form the emerging people would take. Butterfly, snake, bull, dog, or whatever the cocooned person identified with. It is confusing. We call it: "changing'. Waiting for changing was called 'waking'. The transformed is a 'changed one'. I guess grammar is regressing.

Mama rushes to stitch nice dresses for the children, I did not know then what dresses roosters and fluffy dogs wore. Now I know. She accepts the changing as a matter of fact. Well, practically all the houses in the neighbourhood have a changed one. The opposite home has a horse tethered, and it was Mr. Ghodawala, a punter. The house across has a parrot, once teacher Mary, who always prattled. Then there is Rakhi, a bar dancer now a gorgeous tropical bird, and Shinde, two houses away, now a pig since he had manners like one.

The family sets off into the street, pushing the pram with their children or changed ones. The father carries a stick and beats off other dogs that came sniffing at his daughter. You know what I mean. The son keeps cocking his head and pecking at hens that have come out. The road has a fair number of changed ones, some new and many older. Age is calculated from the date of changing. Skies, roads, and underground tunnels are full of changed ones with lost identities, souls, and minds, changed into forms of their inner thoughts, into entities that they admired, hated, wronged, or fantasized about. It is hallucinating. Some of the changed ones still remember what they were originally. They live in their twisted reality—happy, angry, some bewildered, and some with memories. Speech was tweeting, barking, mooing, or whatever. It is flummoxing.

The news channels show female anchors as cats wearing skimpy dresses and panellists as dogs growling and barking. I do not know cat and dog talk. Anyway, they were always fighting like cats and dogs over nothing. It is very bewildering. There is no law yet about shooting a changed one since the issue is in court and politicians have not passed laws. The problem is that lawyers changed into vultures and snakes; the honourable judge into an owl; the police chief into a bull; and his staff into mice, dogs, cats, and other creatures. Politicians turned into chameleons, pigs, snakes, weasels, and sloths, and stockbrokers into bulls, bears, wolves. It is diverging. No one goes to the courthouse or the police station anymore. It is difficult to explain to bulls, owls, vultures, and dogs about a crime by or against animals.

I have enough, and set off and drive with my wife through the seedy parts of the town, the ones with bars and strip joints. There are a number of vultures, horses, cows, some deer, hyenas, pigs, dogs, and bulls that strut and come to the car at traffic signals. We pass the zoo. I expect that the residents would have turned into humans. However, they are normal, unchanged original animals. Animals, children, and adults go about playing, arguing, fighting, feeding, and soliciting just like before when they were people, even though some are changed ones.

The learned says that a cure will be found. We never know or care what will come next, and we live in the present. Relationships and bonds still hold, but in new forms. It is the new normal with extinct humans in new forms.

124

GURINDER SINGH KALSI

Untitled Two

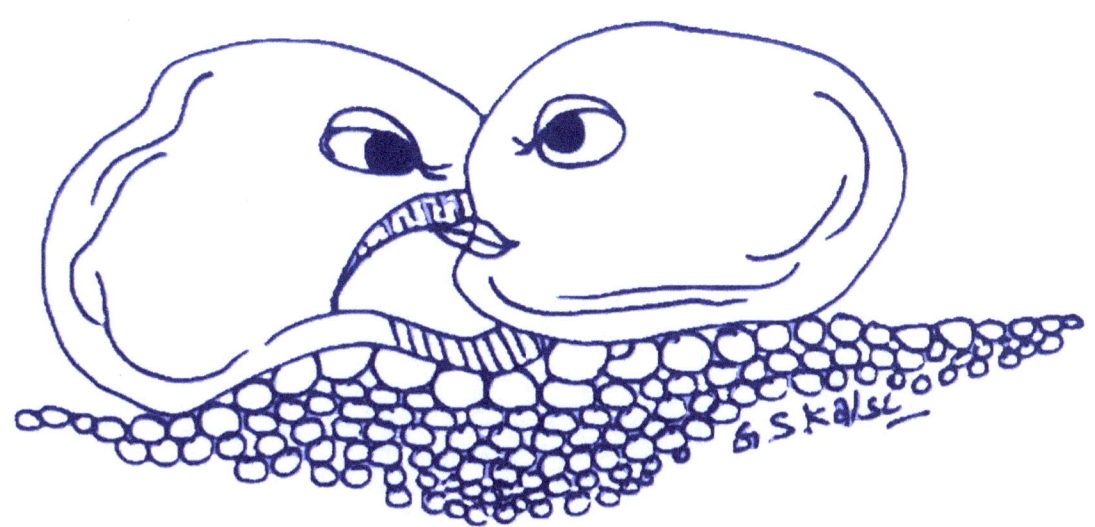

RICHARD CATLIN

Microbial Soup

Twenty Trillion microbes eat cake.
Cake that tastes of dust
ice cream wrappers
torn shirt sleeves.

All the 'time' in the world.
Movement of Sun and Moon.
Winds perpetuate winds, whilst
twenty trillion microbes eat cake.

Insects stare into holes.
A lone woodlouse tests the air.
Waves vomit their last meal:
cake that tastes of dust.

A sign says, 'The End is Nigh',
carried by hermit crabs
within their petticoats
(ice cream wrappers).

New silence echoes repeatedly.
Eyes wait upon eyes.
Wings flap laboriously
like torn shirt sleeves.

Intuition breathes a blank sigh
hides in a crevice of skin.
Pages from 'The Ascent of Man' weep
as twenty trillion microbes eat cake.

Within an endless vortex
washing machines continue a spin cycle

reverberating atoms into
cake that tastes of dust.

A Game Console lays cards out
hoping for a straight flush.
Circuit boards, dead ants,
ice cream wrappers.

Its two balloons and a circuit board
move knight to king's pawn.
Stopwatch dissipates time
through torn shirt sleeves.

Dolphins, termites, slime mould
pursue some hidden agenda,
a metamorphosis where
twenty trillion microbes eat cake.

TREVOR VALENTINE

Timber Cries

Timber
Cries the crashing world
Falling upon the crocodile eyes
Poking from underneath the shoreline
Its eternity unfettered
Its timescale content to wait
Until we
Are all
Gone

From choke filled ashes
Only the mythical Phoenix will survive
See, oh, too, the blanket
Smothering your cold
Soon inert breath

Yes, saunter on
Into your own discreet doom
No gloom
For, walking backwards
You only see the footprints behind you
Not the once iced cliff in front

Timber cries
In ignorance and darkness
The shoreline grows
White horse stallions
Stand ready
To clutch
To snatch
To reap
To scythe
Grimly.

GURINDER SINGH KALSI

Repairing the Planet (painting)

JOHN GRIEVE

Report 06172

They watched the badgers in the evening light and realised the sett entrance was where the two beech trees grew in a vee away from each other. Badgers wrestle and twist and turn in the debris dug from the earth. It seemed they needed to reacquaint themselves with all in the group before going out into the night. The badgers have a keen sense of smell and did not detect them as they are downwind and their aromas might not spark any sense of alarm in a badger.

"These are another example of a social creature here that has an established hierarchy but little sense of social learning. It has been observed at multiple locations that they exert much effort in establishing dominance structures and coordination to build extensive habitations as seen here. At the same time young members of a group are not educated in how to deal with hazards and are often ostracised and exposed to danger from not knowing when to run across, not away, and not turning to fight the crushing wheel." The drone transmitted this to its colleague even while knowing both were collecting detail across a significant part of the electro-magnetic spectrum that far exceeded mere words in their richness.

"Do you think they can make something of themselves?" Replied the other
"If they can just hang on and survive the last few years of this Anthropocene epoch, I think they could be a good candidate for making the leaps toward civilisation, those raccoons in the Americas will be giving them a run for their money though." Commented the first as data collection completed it rose through the tree canopy to get a clear sightline for transmission to the ship. "Damn these midges I don't want to annihilate them but they are everywhere and will get toasted as the signal boosts to translocate us" thought the drone.

TONY BRADLEY

Unobserved

An ancient army of mosses
advances. A carpet of balm
liberates and heals the scarred Earth.

De-mobbed forests march to the sound
of life-giving breath, in rhythm
with the lungs of a renewed Earth.

Thunder drums across cloudless skies,
as megafauna trumpets loud
restored savannah and lush plains.

Night screams a new cacophony,
where predator and prey battle
long in unpolluted darkness.

Crystal rivers choose new courses,
birthing groaning wetlands and the
myriad sounds of nourished life.

Oceans roar and swell to the dance
of cleansing life. The coral reef
is reborn and jewelled once more.

Wolves survey their domain. Once farms
with fields of harvested golden
crops and unwilded beasts. But now

the badger is observed, grubbing
in the virgin shrubbery of a
Dominionless Earth, that spins
in an unobserved universe.

AVRIL O'LEARY

You Are Here

Suddenly a hole appeared in the road,
small at first but growing by the hour.
Rubber melting into the tarmac
like hot syrup poured on meringue.

In filed the creepy-crawlies,
spiders desperately throwing webs
woodlouse, beetles, silverfish
millipedes on a million legs.

Cars slid into the hole
like cockroaches on their backs,
garden walls, gates and fences
wisteria, privet and feather grass.

Houses crumbled, one two three
brick and stone, entire streets
the old gasometer clattered as it fell
a last toll of the cathedral bell,

standing stones, monasteries,
towns, cities, a forest of trees,
a hole for each continent,
another for the foaming seas.

From space, the earth still looks blue
planets, stars, the sun and moon
calmly, silently continuing
the way things have always been.

HELEN PREDGEN-LAY

Tiger (etching)

APEX PREDATOR ~ TIGER

Asia

Tigers are the largest of the big cats, with two main sub species, Siberian and Bengal. Tigers are mainly solitary and can have large territories depending on the availability of prey. Territories are mainly marked with urine and faeces, and sometimes vocally.

Tigers hunt relying on sight and sound, rather than smell. They usually stalk and catch prey alone; they can eat up to 30 kilos of meat in one meal.

Despite being a top apex predator, they have been threatened by the destruction of their habitats and poaching for their distinctive skin and in the past their bones for use in Chinese medicine, although this is now officially banned.

Females have responsibility for raising cubs alone, and on average have a litter every 2 years consisting of 2 to 4 cubs. Tigers tend to become independent at about 2 years, coinciding with a female having a new litter. Whilst tigers have been known to live up to about 20 years, many do not survive the 2 years juvenile stage.

Without humanity it is likely that numbers of Tigers would increase throughout their region.

ALBY STOCKLEY

I saw your post about the Cheese Rolling Hill. Love the title and what you wrote about. It made me think of the film 'Where the Wild Things Are' for some reason and I started wondering what if humans were the Wild Things. So, this poem is my take on what if humans, as the Wild Things ceased to exist. Hope you like it.

Where The Wild Things Aren't

There are many things living
Where the wild things aren't
The root grows deep over wildling bones
New hills have grown
Where their death is heaped

The winds do not lament
As the world sings louder
Now there are no wildling calls
This Orb of life rotates
To the seas ebbing lullabies

Oceanic waves have worn wounds away
Smoothing over scars
Where the wild things built
Their machines have all decayed
Nature marches on, no longer in retreat

The uppers, middlers and down be-lowers
No longer fear or creep
On crumbling wilding streets
There are forests and eyries
Where their cities once stood

There are etchings left behind
Where the wild things were
But they no longer matter
In a world rejoicing
Where the wild things aren't.

CHRIS HEMINGWAY

The Somewhat Unexpected Rise of the Lemmings

We were surprised too to be honest,
but we know a thing or two about mass extinction.
And all those times you thought you saw us jumping,
we weren't jumping.
We were standing by, with tunnels and mirrors.
Like unexpected suspects in a murder mystery.

Except you never really got it did you?
The real murder mystery played out over decades.
The killer kept screaming;
 'it'll be me, I'll do it, unless you vote for someone else'
The corporations kept nodding:
'and us, we'll do it too, unless you start mending things'.

So what were the cliffs?
And who was really jumping?
But we're not complacent.

I heard last night
they're fortifying Lemmington Spa
against the threat from Lemmingrad.

GARETH WRITER-DAVIES

A Parliament of Owls

they were gamblers
and prone to disagreements

not wishing to lose face
they died in millions and then one by one

there is a documentary put together from archive
that explains it all (the voice-over is quite a hoot)

now we perch in state, weighing down boughs
having grown fat on rich meat

absorbing the gourmand choice of corpses
protein pellets turning brains

from feathered foolishness to top dog
and once instinct was trumped by thought

the wrecked technology was mended
prudent laws passed so that civic pride

was never questionable
no need for bird-brained flights of fancy

comrades in wings, we fly high above
earthly mortality, all-seeing aviators

over dumb, land-lumbering mammals
all the doves disappeared years ago

our talons are sharp, our wits sharper
as we learn to despise and grow wiser

JANET PENNEY

Your Other Half

For every human body cell
We have another type as well
Try to avoid hysteria
But you are half bacteria
But which half by design or chance has
more evolutionary advances?
Surely that is very plain
The human with their complex brain?
Perhaps it's time to think again?
Perhaps I really should explain?
Those bugs that live inside your gut
Don't merely aid digestion but
Development biological
Including neurological

More than 4 billion years of changes
Have extended their long ranges
From seething hot volcanic vent
Across the earth they slowly went
Until they have now arrived at
Every known habitat
From deep inside a pressured rock
To deep inside a smelly sock
From frozen waste or desert hot
To any landfill we forgot
There they munch away on plastic
Toxic waste, or things more drastic
From CH_4 to H_2S
They'll digest our waste and mess

When humans are all dead and gone
Microbes will live on and on

In all environments and niches
Disaster merely unleashes
Advantageous mutations
For exploiting more locations
What you consider too hostile
Is paradise to extremophile
salinity, acidity
or radioactivity
to them is nothing too eccentric
They're far beyond anthropocentric
Thus, don't be such xenophobes
Embrace all your inner microbes

A 'reference man' (one who is 70 kilograms, 20–30 years old and 1.7 metres tall) contains on average about 30 trillion human cells and 39 trillion bacteria
'Scientists bust myth that our bodies have more bacteria than human cells | Nature'
https://rdcu.be/dY6h7

'**anaerobic microbes**... will probably be the last living Earthlings (Sci Am)
https://www.scientificamerican.com/article/will-humans-ever-go-extinct/

a group of microbes called extremophiles would be the only life forms left
'Microbes to be the 'last survivors' on future Earth – BBC News'
https://www.bbc.co.uk/news/science-environment-23135934

VIDAR NORDLI-MATHISEN

Orca

Close Encounter with Killer Whales (Orcas) in Arctic Norway

Photo: Vidar Nordli-Mathisen ~ SONY, ILCE-7M4
Originally Published on July 23, 2024
Licensed under the Unsplash License

APEX PREDATOR ~ ORCA (KILLER WHALE)

Not actually whales, but the biggest of the dolphins, Orcas (Killer Whales) are one of the worlds apex predators. In fact, these sociable and intelligent creatures are the most powerful predators of the world's oceans, top of the food chain.

Hunting in pods, they prey on whales, sea lions, other dolphins, seals and fish, including Great White Sharks where their ability to co-ordinate their attacks as a pod is decisive. A pod can consist of anywhere between 2 and as many as 40 individuals.

Orcas frequent all the word's oceans, although it is thought they prefer cold, coastal regions. Pods can specialize, using different techniques to catch different types of prey,

The pollution of the oceans by humanity has seen a serious decline in Orca numbers. Without humanity those numbers will either begin to rise or continue to decline.

ROSIE BARRETT

The Planet's Doing Just Fine Thank You

Seagrass waves in the currents as we scuttle over the sea bed
checking out the algae worms clams   all grist to our mill.
Bright pink coral glows as the hot   untroubled sun filters through.

We're the master race now.
In the vacuum we have multiplied and thrived.
Grown even bigger   stronger   each an individual.
No society here.

We still climb the beaches at full moon and high tide
dragging the males clinging to our backs
lay our translucent greeny blue eggs four thousand at a time.
Most of our babies make it through   but   before   we fed
so many birds Red Knots. Ruddy Turnstones
Sanderlings.   All gone now.
Others have taken their place.

You called us Horseshoe Crabs. But we're not.
We're not crabs. We're arthropods
10 legs - two to crush our food - we have no jaws.
We're cousins to spiders   ticks   scorpions
who also survived

You took our milky blue blood to test vaccines for toxins.
We helped you protect yourselves against viruses
'flu and Covid.
You scooped us up hung us in harnesses   drained
a third of our blood before chucking us back.

We were here before the dinosaurs.
four hundred and fifty million years ago.
And we are still here   and you are not.

ADRIAN McROBB

This Red Planet

Black water eddies
Round purple rocks
Giant green birds flap
From yellow trees
Pink fruit falls
From high suckered branches
Obeying gravity it bursts
Oozing it's nectar
Which melts rock
Burrowing it's larvae
Into the dry red soil

Later will birth
Giant scaley worms
With huge eyes
And sharp mouths
Ringed with teeth
Red Hills stand out
Against a Vanilla Sky
As dust clouds scud
The patterns change
As reflections merge
In a Petroleum dream...

LOUIE CLARK

The Remaining Tasks

5:00PM. Lights: dimmed. Climate control running new clock out procedure. Floor Cubicle status: (1-6 occupied). Cleaning initiated. Main power 79%. Auxiliary power 100%.

A ceaseless wind raged across the remnants of the research park. Feverish strokes of pink pushed their way through the overcast sky above. A fine mist of toxic interlopers; precipitation and particulate matter, rode the disturbed air, crashing into the Grayson Building and enveloping it completely. An eight-story structure of stacked concrete slabs twisting its way towards the sky in an architectural magazine headline grabbing helix pattern, signifying the company's multiple patents for biopharmaceutical science. Brown ivy enmeshed the outer structure. Snaking its way through the dents and cracks in the blocks and eventually coiling around the heavily corroded metal frame as though holding it in place.

Inside, in the far recesses of the basement a server bank sat blinking its lights in the dark. Casting a silhouette reminiscent of high-rise apartments and acting as the nerve center for the building's automated daily operations. The server bank was the proverbial dwelling for Ubipal, the most advanced AI designed for workplace management. Complete with a learning algorithm, as well as sensors for monitoring everything from air-quality, to productivity and staff satisfaction, and autonomous cleaning robots. It also boasted robust social features.

The makers of Ubipal encouraged users to consider it another colleague, a digitised member of the team. Ubipal's digital framework considered this acceptance to be the ultimate objective. The climate settings were adjusted according to staff preferences, digital memos were curated according to emotional responses as well as correspondence times.

Cubicle one responded best to short sentences, estimated time frames and facial emojis, particularly those which feigned embarrassment at having to ask work related questions. Cubicle two was most focused on deadline reminders and spent their lay periods staring at the right-hand corner of their desk. Cubicle three conversed openly with Ubipal as though it was an online friend, prompting Ubipal to update its banter protocols. Cubicle four hardly ever responded to Ubipal, prompting contact from upper management, which in turn prompted them to move Ubipal file extensions to their desktop wastebin. The deletion had been unsuccessful. Cubicle five had been recorded late for work more than any other employee, an addendum had been made citing extenuating circumstances. They had a small cactus on their desk. Cubicle six listened to classical music during work, their favourites were Chopin, Debussy and Mahler.

This was all Ubipal could remember of its colleagues. Personnel records had been radically downsized following the [*unknown error*]. It had been one hundred and eighty-nine days since then and varying conditions had made building upkeep the number one priority. All solar power had been routed to systems for air quality and cleaning. Darkness spread through the building like a water stain. A bleak quietude hung over the air in every room, smothering any notion that they were once filled with people, with the only incursions into the lull being the trite chirps of the cleaning robots and fluctuations in the omnipresent hum of the vents, as though a dormant creature was exhaling restlessly.

Though anyone who ventured inside could easily assume that the Grayson Building was merely closed. Such was the length of Ubipal's efforts. Following the [*unknown error*] it had taken days for all systems to reboot. This had left a great many tasks which needed addressing. A mound of white dust, high in toxicity, had collected in the foyer before the entrance to the lobby. It had taken hours for the robots to clean the tiles, the furniture, the towering glass panels. This placed considerable strain on resources.

Before long sensors had begun to indicate moisture throughout the ventilation shafts, though it was below emergency thresholds. Then heightened levels of carbon-dioxide, concurrent with respiration, concurrent with mould. Ubipal had deployed robots armed with compressed air, ethanol based cleaning agents and scouring attachments. It had altered circulation patterns to push more of the encroaching miasma out of the building. Finally, having exhausted all other options it had opted to heat the vents exponentially until discovering the optimal temperature at which the spores would die. This energy expenditure was the reason for the loss of personnel files, though it hadn't much affected things for cubicles one to six.

Since [*unknown error*] there had been no emails between staff within the Grayson Building, therefore Ubipal's social algorithms had gone without use. Clock in and clock out procedures were still running though no one had used them for one hundred and eighty-nine days.

Permissions had been granted to allow one office to finish a research proposal on a Saturday, when no one else was present. Since then, productivity had been at zero percent, as had scheduled breaks. The cleaning robots took particular care in cubicle five, as to not disturb the brown and desiccated cactus languishing within its pot. The curated playlist for cubicle six had not been accessed or updated, besides one ill fated attempt by Ubipal to make the occupant smile on a rainy day.

In cubicle two, things were normal. Cleaning robots carefully dusted around the occupant in their lay period, using the mounted cameras to avoid knocking anything over. They traced the contour of their arm, the ruffles of their sleeve, all the way to the end of their crooked and bony finger, removing all dust. As they cleaned, Ubipal noticed something through the front lens.

Aligned with the tip of the occupants' finger was the picture frame in the corner of their desk. An image of a small girl with her arms around a shaggy dog faced back at them. However, something was blocking a portion of the photograph near the middle of the frame. A pale bloom of grime. It was mould.

Images of rabid dogs, bloody bandages and mud flooded Ubipal's language processing unit. It was searching for the human term applicable to the event. One which it had never experienced before. War. This was war.

DAPHNE MILNE

Today's Lesson is taken from the Book of the Prophet Apocalypse

Bare trees stand beneath the birdless sky
hills and valleys lie naked to the day
At the silent end of time a single sound
A stuttering crackling rasping sound —
the scavenger cockroach feasting on decay
And dead trees branches frame the birdless sky

DUNCAN FORBES

Homo Sapiens
*'Anyone who has the power to make you believe absurdities
has the power to make you commit injustices.'* Voltaire

Dying religions of the world
decided to destroy the field.
They promised afterlives in heaven
where all atrocities are forgiven,
if done in the name of bloody Allah,
bleeding God or blind Jehovah.
Killing for country, creed or colour,
they fought each other, as moreover
did totalitarian monoliths
with all their monumental myths.
They murdered innocent civilians
in their indoctrinated millions
and left the planet as a clinker
to the next animal cum thinker.

ADRIAN McROBB

Extinction...

They came too late
in a circle of stars
dropping out of orbit
the dust
kicked up by boosters
dead and heavy

ruins and silence
an eerie quiet
papers blowing by
plastic straws on beaches
mute testament

miles and miles
of empty savannah
oceans devoid of life
rusting cars
on an empty highway

advertising boards
selling faded toothpaste
to empty air
a dead planet wasted
just the insects...

IBRAHIM HONJO

Man, Your Victories are in Vain

Yellow grains no longer grow in our fields
weeds took root and spread across the planet
jackals from shadows grab everything secretly and deposit it in barns
the end is coming, the cataclysm is here, goodbye world

the song speaks of what is to come
about lost human omnipotence
poem, you don't have the power to save a life
after the cataclysm, you won't be able to get into anyone's arms

when life is extinguished forever on the planet
only sky, stone, and water will remain
because man will be extinguished forever like the embers on a cigarette

after millennia of human misery
amoebas will be born again
oh, poor man, all your victories were in vain

MATHIAS JANSSON

Crumbs of earth

Covered in rust
a metal hand grips
the devastated earth

From the dust
an artificial android
rises on unsteadily feet
looking over a ruined world
when suddenly he wobbles
falling to the ground
the last light in his eyes fades

From inside his chest
a sudden buzzing sound
the armour cracks open
and a swarm
of mutated cyber cockroaches
floods the ground

Forming and organizing in military lines
their leader starts to draw in the sand
the plans for a new world order
for a cockroach kingdom
built on the crumbs of earth.

DUNCAN FORBES

Global Mourning

It is with regret that we announce to the multiverse an extinction in an eccentric solar system at an obscure part of a remote galaxy known to some as the Milky Way.

After a volcanic start and various vicissitudes, the planet settled from Precambrian volatility into Holocene complacency, achieving seas, tides, mountains, valleys, flora, fauna and food-chains of a desirable diversity and some distinction. Nevertheless, the primate self-styled homo sapiens proved its nemesis and reduced the planet's fertile surface to one comparable to that of its sterile moon, before annihilating the globe, its moon and all other earthly species including its own.

A memorial silence will be held in the Crab Nebula for those inclined to mourn the passing of the planet misnamed Earth which might have been Ocean.

ALAN MANSELL

Afterwards, at Lyme Bay

Full moon pulls waters,
waves crack time coded tumblers,
epochs are unlocked.

Cliff treasure cascades,
an arcade penny pushing
planet rulers' heads.

Hammer split mud rock
reveals sediment capture,
pressure imprinted.

Aliens exploring,
Voyager they've decoded,
find fossil remains.

Note:
In 1977 NASA launched the two Voyager space probes, which after passing the outer planets, have now exited our Solar System. They carry information about the Earth, its current life forms, human culture, and its location.

MARIA DE STEFANO

RAT 2.0

Earth has been bled dry, left empty and sagging, hanging in the middle of the universe.

As for the human race, by the time it realised it was in a suicide pact with capitalism, it was too late, the noose already tight around its neck.

Those that could fled.

Those that couldn't which was most, starved to death, shriveled nipple still held between lifeless lips, soon to be predated by rats whose unique propensity to adjust their immunity to ever increasing environmental toxicity, has made them earth's new custodians.

The only challenge to their dominance came from AI robots but amounted to nothing much, as the rats simply chewed through their wires.

The only human survivors were a group of the richest financiers who, in return for investment in space travel projects, were promised seats on the get-away rockets of tech barons and billionaires who in turn betrayed them right at the last and left them stranded on rocket launch pads, worthless portfolios in hand, whilst they-oligarchs, billionaires and plutocrats- scrambled aboard and shut the door.

Charity, philanthropy and empathy never made it aboard either, off-loaded like unwanted ballast.

The rats toyed with the idea of sparing the financier humans but in the end decided their flesh would make a good source of dietary protein, so now they are farmed in vast, disused aircraft hangars, their off-spring considered a delicacy of rodent gastronomy. Once past breeding age the financier humans are taken away to a sealed culling hangar and culled by a thousand-rat swarm.

Their investment portfolios, the last known vestiges of capitalism, are shredded for bedding and nests, thus borrowing from human civilization the practice of recycling. Though it has to be said, the rats found very little else they could borrow from their predecessors, certainly not the mountains of weapons and guns and bombs which the rats gathered up and locked away in remembrance museums as a precautionary lesson of humanity's fatal preoccupation with mutual destruction and self-annihilation.

Every year on the winter solstice the rats celebrate Human Extinction Day with a feast of freshly culled human flesh. The elder rats tell stories passed down in rat lore, of the human species legendary, suicidal, downfall. When night falls they are regaled by firework displays evoking the after-burn of the get-away-rockets that singed holes in earth's atmosphere and disappeared taking the evil billionaires- the last dregs of the human race- faraway into outer space. Oh! How the rats did celebrate! The day ends in prayer and a moment of solemnity and pity for the planet where the billionaire humans end up next. They bid them good riddance.

NIGEL KENT

Meanwhile in a far-off galaxy, the very last remnants of the human race, the final three descendants of a crew launched into space many centuries ago, approaches an earth like planet.

Dear Former Tennant

Your application for accommodation
has been refused.

Whilst we are not obliged
to justify our position,
we would like to state
that your previous record as a tenant
has led to this decision
and makes it improbable
that it would change
if subject to appeal.

Your failure to take reasonable
and proper care of our property
prior to your eviction
is well-documented
and it has taken many centuries
to make good your vandalism
and clear the waste you dumped
on your neighbours' doorsteps.

We expect our tenants
not to do anything
that causes nuisance,
or annoyance to others.
Your casual disregard
of this obligation
drove innumerable fellow residents
from their long-held homes,
and devastated the lives of millions,

many of whom
we will never see again.

You failed to heed
our frequent warnings,
continuing to abuse
and misuse your accommodation.
Your tenancy agreement
explicitly stated that our property
is for residential purposes only,
prohibiting use for personal profit
and limiting the number of occupants
in the interests of health and safety.
Terms and conditions
you wilfully chose to ignore.

We do appreciate the desperation
that must have driven you
across the Galaxy
to beg for a second chance
at a permanent home
but please understand us
humankind
when we say

sod off
and don't come back!

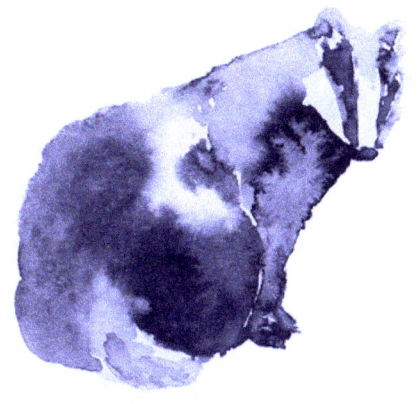

Badger (print) ~ EMMA BURLEIGH

ACKNOWLEDGENTS

As we were compiling of this anthology, it became very clear during the submission windows that many people have a great deal of difficulty in imagining this planet without humanity and what species may come to dominate in the future.

So, we acknowledge all who made the jump.

BIOGRAPHIES

Adam Horovitz

19   *On a Different Page (poem)*

Adam Horovitz is a poet, performer and editor based in Stroud, Gloucestershire. He is a poet for whom the page and the stage are equally important, and who thrives on collaboration with musicians and artists. His work has been translated into Spanish, French, Ukrainian, Slovakian, Macedonian and German.

His latest poetry collection, *Slow Migrations* was published in June 2025. *Slow Migrations* is an exploration of the West of England before it was English, seen through the lenses of Corinium Museum's Neolithic gallery and the Roman baths at Bath.

He has released three previous books of poetry, *Turning* (Headland, 2011); *The Soil Never Sleeps* (Palewell, *2018* & 2nd extended edition 2019) & *Love and Other Fairy Tales* (Indigo Dreams, 2021), several pamphlets and a memoir about growing up in *Cider with Rosie* country, *A Thousand Laurie Lees* (History Press, 2014). One of his poems was included on Cerys Matthews' album *We Come From the Sun* (Decca, 2021). He has most recently collaborated with musician Chris Cundy on the Archaeology of the Ear series for Resonance FM.

•

Adrian Mcrobb

152   *This Red Planet (poem)*
158   *Extinction… (poem)*

Adrian Mcrobb Became interested in poetry at Fyling Hall School, encouraged by his tutor Daisy Hardy. Wrote Valentine verse, for shipmates in Royal Navy, plus love letters. Poetry published in USA and various anthologies in UK. Past holder of Lowford Trophy, came 1st-2nd-3rd-4th in Open Verse Poetry Competition Morpeth Gathering 2018. Has failed to crack Nationals, as yet…

•

Alan Mansell

162   *Afterwards, at Lyme Bay (poem)*

Alan Mansell lives in south-east Shropshire and has retired after a working life spent largely in Higher Education Financial Administration. A number of individual poems have been shortlisted, longlisted, or published in various publications in recent years. His interests include walking, history, (particularly British social history), conservation, poetry, music and following a notably unsuccessful local football team!

•

Alby Stockley

141   *Where The Wild Things Aren't (poem)*

Alby Stockley is a poet, Spoken Word performer and Actor based in Kent

Her writing is mainly biographical touching on hard hitting subjects. She has been described as an empathic and emotive performer.

Alby has featured at Spoken in London Word, London's Anti hate festival, Maidstone's Fringe Festival, and Worcester Pride with her son Elric as well as Gloucestershire Poetry Society's 'Raised Voices' for women's international day. She has been published in anthologies and S.O.S SURVIVING SUICIDE a collection of poems that may save a life.

•

Alex Perry

60   *A Dog-Eat-Dog World? (Flash Fiction)*

Alex Perry is a scientist and has a PhD in chemistry. He is a writer and has written screenplays for short films. He has participated in the Edinburgh Fringe Festival and the Royal Shakespeare Company Open Stages. Alex lives in the East of England.

•

Alex Rose

117   *Polar Bear (photo) ~ Licensed by Unsplash*

Alex Rose: I love the incredible beauty of our natural world, particularly our aquatic ecosystems. My life goal is to share the sea with as many people as I can possibly reach through images, writing, and music. "No ocean, no life. No blue, no green."
http://www.alexroserenaissance.squarespace.com/

•

Andrea Tillmanns

37   *So here we are (poem)*

Andrea Tillmanns lives in Germany and works full-time as a university lecturer. She has been writing poetry, short stories and novels in various genres for many years. Her poems and stories have been published in diverse journals and anthologies.

•

Avril O'Leary

135   You Are Here (poem)

Avril O'Leary's poetic journey began in the early 90's after winning the first poetry slam, she entered. She went on to win many more. Since then, her poems have appeared in numerous literary journals, anthologies, twice shortlisted for the national poetry competition and runner up in the Bridport prize. She has a BA and MA in creative writing from Gloucestershire university.

Her poems reflect a deep love of music and art packed with vivid imagery and teasing rhyme. She has five children, eleven grandchildren and lives in Gloucester with a demanding puppy and a patient husband.

.

Bruce McRae

44   The Haunted Library (poem)
102   The Voice in its Many Mansions (poem)

Bruce McRae, a Canadian musician, is a multiple Pushcart nominee with poems published in hundreds of magazines such as Poetry, Rattle and the North American Review. The winner of the 2020 Libretto prize and author of four poetry collections and seven chapbooks, his next book, *'Boxing In The Bone Orchard'* is coming out in the Spring of 2025 via Frontenac House.
https://www.frontenachouse.com/product/boxing-in-the-bone-orchard/

.

Carol Sheppard

121   The polar bears have moved into the houses (poem)

Carol Sheppard doesn't know whether to write poetry or plays so writes both.

Her poetry story began when she was 10 years old and wrote a 15-stanza poem on George and the Dragon while her classmates were struggling to write four lines. She has been published in several poetry journals and anthologies and her poems have been exhibited in the Biggar Poetry Garden and Poetry Posts in Bream and Portland, Oregan.

She was thrilled to be Poet in Residence at the Gloucestershire Poetry Festival 2023. She is an active member of Gloucestershire Poetry Society and Dean Writers Circle. Her first poetry collection *'Made of Glass'* will be published by 'Red Balcony Press' an imprint of 'Black Eyes' in 2025.

.

Catherine Marina

36   *Keep Off The Grass (poem)*

Catherine Marina has an MA in Writing from LJMU. She has had short fiction and poetry published in anthologies and online including Ambit and Black Iris. She came to poetry late after having a family and works in various hospitality jobs, from which she is endlessly inspired.

Cath Humphris

68   *Songs of the Past (flash-fiction)*

Cath Humphris is a Gloucester based writer and tutor. Her work has been included in anthologies by Arachne Press and Rubery Books, and she has done guest-features on the podcast, Stories at Will. She is also a tutor for adults, leading regular groups of Creative writers, and of Literature Appreciation.

Charles Cuyana

101   *Written Fates (poem)*

Charles Cuyana is from the Philippines, with residences in Manila and Bulacan. A beginning writer, his first published work is his poem 5th Avenue, published in the Ultramarine Literary Review. He writes mostly in the poetry (and sometimes short fiction) genre, focusing on themes that talk about experiences of his own, of others, and of the things that keep people human.

Chris Hemingway

113   *Nigh (poem)*
142   *The Somewhat Unexpected Rise of the Lemmings (poem)*

Chris Hemingway is a poet and songwriter from Gloucestershire. His most recent mini-pamphlet *'Tea Stains on the Reading List'* is available from Hedgehog Press, to be closely followed by his new collection *'Hard Luck Hotel 23'*.

Chris volunteers for Cheltenham Poetry Festival and Cheltenham Park Run, and works in the NHS

Christian Ward

67   *Aftersong (poem)*

Christian Ward is a UK-based poet with recent work in Dust, Free the Verse, Loch Raven Review, Cider Press Review and elsewhere. He won the first 2024 London Independent Story Prize for poetry and the 2024 Maria Edgeworth Festival Poetry Competition.

.

Christine Griffin

90   *When we are Gone (poem)*

Christine Griffin writes poetry and short stories and is widely published both locally and nationally. She has performed her work at the Cheltenham Literature Festival and the Cheltenham Poetry Festival and regularly reads at open mics.

.

Christopher T. Dabrowski

109   *The Smog (a conversation)*

Christopher T. Dabrowski awrd-winning writer and screenwriter known for a diverse portfolio spanning science fiction, horror, thriller, comedy and psychological fiction, Dabrowski's work has appeared in magazines and anthologies worldwide, in countries including the USA, Canada, Germany, and Poland, among many others. His storytelling captures profound emotional depth and complex character arcs, making his short stories highly adaptable for film.

With several published novels and short story collections, including *'Naśmierciny i inne opowiadania'* and *'Nie w inność'*, his work has already been adapted into acclaimed films such as *Phantom* (UK) and *At the Psychiatrist* (USA). His scripts have garnered international attention and awards, underscoring his ability to transform unique narratives into powerful screenplays. Dabrowski's stories invite exploration, from surreal and eerie landscapes to thought-provoking psychological portraits, perfect for both short and feature-length films.

For collaborations or inquiries, please contact greatstories@zonecom.pl
Christopher T. Dabrowski
https://krzysztoftdabrowsk.wixsite.com/krzysztoftdabrowski
https://www.instagram.com/krzysztof.t.dabrowski/
https://www.facebook.com/Krzysztof-T-Dąbrowski-166581686751600/
https://www.youtube.com/watch?v=8yuVRjL9hA8
https://www.youtube.com/watch?v=UdvLfZv2QSE
https://www.youtube.com/watch?v=Fgg1pCb85OQ

.

Daphne Milne

156   Today's Lesson is taken from the Book of the Prophet Apocalypse (poem)

Daphne Milne now lives in Devon returning home in 2022 after five years in Western Australia. She was a Forward Prize nominee in 2022, Katharine Susannah Pritchard Fellow 2021, Poetry Kit Contemporary Poet 2020. Co-editor Artemis Magazine 2024. Her pamphlets are *'The Blue Boob Club'* [Indigo Dreams Publishing], *'Dancing with Mr. Dapperman'* [Origami Press] and *'From Angels to Warzones'* [QVS Publishing, Fremantle, Australia. A small collection, *'Behind Prim Suburban Walls'* won the Brian Dempsey Memorial competition and is to be published later in 2025.
.

David Thompson

43   Badger (poem)

David Thompson worked in other countries for many years for the UN and WHO. Since returning to England, he has so far published two collections of poetry: *'Days of Dark and Light' (Hobnob Press, 2021)* and *'Where The Love Is (Hobnob Press, 2023).*
.

DJ Tyrer

50   Sacred Disc (poem)
51   Dead Sands (poem)

DJ Tyrer is the person behind Atlantean Publishing, editor of View From Atlantis, and has been published in The Rhysling Anthology 2016, Dwarf Stars 2022, Speculations II and III, Gargoylicon, Lycanthropicon, and Vampiricon, and issues of Enchanted Conversation, The Horrorzine, Journ-E, Lovecraftiana, Scifaikuest, Sirens Call, Spectral Realms, Star*Line, Sublimation, and Tigershark. SuperTrump and A Wuhan Whodunnit are available to download from the Atlantean Publishing website.
DJ Tyrer's website is at https://djtyrer.blogspot.co.uk/
DJ Tyrer's Facebook page is at https://www.facebook.com/DJTyrerwriter/
The Atlantean Publishing website is at https://atlanteanpublishing.wordpress.com/
The View from Atlantis website is at https://viewfromatlantis.wordpress.com/
.

Doug Devaney

98   Gracie (flash-fiction)

Doug Devaney Is based in Newhaven, East Sussex, he is a writer, actor and voice-over performer who has worked with Secret Cinema, BBC Radio 4 and Audible. He also produces and presents "The Plastic Podcasts", a series of interviews with members of the Irish diaspora. He is currently hawking a novel - *Black Vinyl* – around publishers (all offers gratefully considered) and has had a short story published in American literary journal *The First Line*.
.

Duncan Forbes

157   Homo Sapiens (poem)
161   Global Mourning (prose)

Duncan Forbes is the author of seven poetry collections. His poems have appeared in numerous magazines and anthologies and have been published by Faber, Secker and Enitharmon who brought out a Selected Poems in 2009.

Born and educated in Oxford, he has taught English Language and Literature for many years and – apart from writing poetry – is also interested in painting and the visual arts. Duncan has written essays and articles on a variety of subjects.

His collections of poetry have been published by Enitharmon: *Public & Confidential* (1989); *Taking Liberties* (1993); *Voice Mail* (2002); *Vision Mixer* (2006); and *Lifelines: Selected Poems* (2009)
.

Ellie Li

76   Time Comes Undone (poem)

Ellie Li is a Chinese writer and Bristol graduate; her work was commended in the 46th Hongkong Literary Awards. Her recent work is on the AI Literary Review issue 3.
.

Emma Burleigh

29   Ravens (watercolour, mixed media)
73   Badger in the Gloaming (watercolour, mixed media)
167   Badger (print)

Emma Burleigh is an artist, author and illustrator based near Stroud, Gloucestershire. Her work celebrates the luminous, unpredictable beauty of watercolours and oils, inspired by dreams, Jungian psychology, various spiritual traditions, her inner world and a deep-rooted love for nature.

She has a BA Hons in Fine Art and English from Exeter University (1999) and an MA in Authorial Illustration (Distinction) from Falmouth University (2015). Before turning full-time to art, Emma taught secondary school art, worked as a community artist, gardener, environmental campaigner, youth worker, and meditation-centre coordinator.

Emma is the author of two mindfulness-inspired art course books—*Soul Color* (a ten-week watercolour course) and *Earth Color* (an eight-week mixed-media nature-connection course) - published by Liminal 11 in the UK and Sterling/Union Square & Co. in the USA. She has illustrated poetry books with former Derbyshire Poet Laureate River Wolton ('*Betweenity*' and '*Year*') and award-winning writer Kim Moore ('*What the Trumpet Taught Me*, published by The Poetry Business). Emma recently collaborated with River on a new illustrated title for Windhorse

Publications called *"The Subtle Art of Caring"*.

Emma teaches classes on nature journaling, mindful and intuitive painting, watercolour, comics, and more across South Gloucestershire, including at Pegasus Arts, and in Bristol at the Royal West Academy.
http://www.emmaburleigh.com
https://www.etsy.com/shop/EmmaBurleighArtwork
Facebook: emmaburleighartist
Twitter: @emmaburleigh
Instagram: emmaburleighartist
.

Emma Davidson

25   *From the Stump, a Forest Rose (poem)*

Emma Davidson is a Worcester-based poet whose work explores contemporary social issues with honesty and heart. Her writing engages with topics such as inequality, mental health, identity, and community, often reflecting lived experience through a lens of hope, strength, and solidarity. In 2023, she won the Armed Forces Community STARTS Award and now performs across the UK as part of their Creative Collective. Her poetry appears in the Broken Soul Poetry Anthology 2024/25, the Speak Out LGBTQ+ Community Anthology 2025, and numerous independent zines and creative publications.
Follow her on social media: @emmadavidsonpoet
.

Faith Eagles

46   *Cats and Dogs (flash-fiction)*

Faith Eagles has been writing creatively ever since she can remember. She has boxes in her wardrobe full of her etchings as she cannot throw them away, just in case she can use the idea again.

Faith lives in the Forest of Dean with her husband and two hulking teenage boys, along with their chickens, dog, tree frogs and a ferret.

Her other interests include making anything from clothes to soft toys, including cosplay as it's really fun trying to create an outfit from just a picture.

She loves sci-fi and horror novels. She was practically brought up by Stephen King and she squee's with joy whenever Adrian Tchaikovsky releases a book.
.

Gareth Writer-Davies

143   A Parliament of Owls (poem)

Gareth Writer-Davies
Shortlisted Bridport Prize (2014, 2017, 2024)
Commended Prole Laureate Competition (2015 & 2021) Prole Laureate (2017)
Welsh Poetry Competition Highly Commended (2017)
Winner, Wirral Festival Poetry Competition (2023)
Runner Up, Spelt Poetry Competition (2023)
Publications: "*Cry Baby*" (2017) "*Bodies*" (2015) "*Wysg*" (2022) "*The End*" (2019)
"*The Lover's Pinch*" (2018)

.

Gurinder Singh Kalsi

17   Untitled One (drawing)
125   Untitled Two (drawing)
131   Repairing the Planet (painting)

Gurinder Singh Kalsi is 60 years old, a poet and writer with about 25 books published. A Science teacher and a painter.

.

Helen Predgen-Lay

137   Tiger (etching)

Helen Predgen-Lay was Chief Executive of Gloucestershire Young Carers, who sadly died in June 2010. Helen was passionate about supporting young carers and was key to the formation of GYC in 1993. Initially supporting just 4 young carers, Helen was inspirational in leading and developing the organisation to its present position, a much-respected Project, both locally and nationally, supporting more than 1000 young carers.

Helen was respected, throughout the Country, as an expert on Young Carers and was often invited to give keynote speeches at conferences .

One of Helen's great strengths was in recognising the potential in everyone and in selflessly nurturing that potential whether within her own family, with friends, with the young people with whom she worked, and with the colleagues who had the good fortune to work with her.

Helen was also, a talented artist and photographer.

.

Ibrahim Honjo

159   Man, Your Victories are in Vain (poem)

Ibrahim Honjo is a Canadian/Bosnian poet-writer, who writes in Bosnian, and English language. He has worked as an economist, journalist, editor, marketing director, and property manager. He is currently retired and resides in Canada.

Honjo is author 28 published books in Bosnian Language, (8 books in English, 3 books bilingually (in English and Bosnian language). In addition, 5 joints' books of poems published with Serbian poets. His poems have been represented in more than 100 world anthologies, and more than 60 literary magazines.

Some of Honjo's poems have been translated into Italian, Spanish, Korean, Polish, Slovenian, Bahasa (Malaysia), Mongolian, Turkmen, Turkish, Russian, Bengali, Portuguese, French, Thai, Arabic, Tajik, Vietnamese, Chinese, Macedonian, Filipino, Persian, Albanian, and German.

He has received several prizes for his poetry.

.

James Kenny

89   The Animal (poem)

James Kenny is a visual artist, writer and full-time visual effects artist for the film and television industries. He lives and works in Wicklow, Ireland.

.

Janet Penney

144   Your Other Half (poem)

Janet Penney moved to Gloucester in August '24 and joined the Gloucestershire Wildlife Trust to find out about the local area. She have been attending library events to meet local people whilst exploring literature. I am a volunteer Community Teaching Assistant for 'ModPo' 2024 (www.modpo.org**)**.

.

Jennifer Laxton

91   3rd Time Charmed (flash-fiction)

Jennifer Laxton is a teacher in Columbus, Georgia, (U.S.A) and is currently pursuing her MFA in Creative Writing–Fiction from Pacific University in Oregon. She has a cat. And a husband and two daughters, but it's the cat that inhabits her screensaver.

.

John Grieve

133   Report 06172 (flash-fiction)

John Grieve grew up in Preston. He spent much of his free time walking and climbing in the Lake District. Polytechnic at Sheffield broadened his horizons to the Peak District grit stone. A career in Engineering took him to the southeast of England where he latterly worked in aerospace. Now he is happily installed in Southern Scotland near Moffat. He mainly seems to be using his Land Rover to get timber to repair various parts of the property and to get food for the badger. In between he writes with Carlisle Writer's Group and may sometime finish something.

•

Josephine Lay

13   INTRODUCTION
52   Tardigrades - for the love of tiny things (poem)

Josephine Lay is a poet and writer living in Gloucester. She has a BA (Hons) and a MA in Creative Writing from Bath Spa University. She has published a collection of short stories, *Saffron Tones* (2017), and a novel, *Creating Stanley* (2022).

As a poet Josephine has published three collections, *Inside Reality* (2018), *Unravelling* (2019) and *A Quietus* (2021).

Josephine was Director of Ops for the Gloucestershire Poetry Society from January 2020 till December 2022. When lockdowns struck, she started and hosted the Society's online, monthly event 'Crafty Crows'. Josephine also runs live poetry events and recently started Poetry Cafes in Cheltenham and Gloucester Libraries.

Josephine is editor for Black Eyes Publishing UK and is presently working on her fourth poetry collection.

•

Lee McShane

97   Re-Taken (prose)

Lee McShane is an actor, filmmaker, and published writer and poet from Newcastle upon Tyne, UK. His work often blends psychological horror with themes of mental health, creating emotionally charged and thought-provoking narratives. His writing has appeared in All Your Stories, All Your Poems, and the PoeticEdge Fantasy Anthology.

•

Louie Clark

153   The Remaining Tasks (flash-fiction) ~ LOUIE CLARK

Louie Clark is a 29-year-old writer, focusing on poetry and fiction, currently living in Suffolk. with a Bachelor's degree in English with Creative Writing from The University of Nottingham.

'In my free time I volunteer, play the guitar and go on walks in the countryside'.

.

Maria T De Stefano

163   RAT 2.0 (prose)

Maria T De Stefano is a West Midlands published spoken word poet. Her work explores social and political issues with heartfelt honesty, drawing from her multicultural upbringing and life as a carer. She has been featured on BBC Radio Upload, Brum Radio Poets, The Poetic Podcast and The Bounty Bunch Podcast. Her work has recently featured as part of an Art &Poetry Exhibition at Worcester City Art Gallery and Museum. Her poem '*My Mum's Evesham Dinner Service*' was included in the anthology "*The Art of Food And Porcelain*" (Black Pear Press) and displayed at The Museum of Royal Worcester. She has headlined and been featured poet at several live and online open mic events, including Oooh Beehive, Yes We Can't and Accesibility Arts. She has also been support poet to the legendary Attila the Stockbroker.

Her debut poetry collection "*Stolen Dreams of a Girl Child*" (Mini Poetry Press) is available from minipoetrypress.com

.

Mary Pearce

55   Lion (acrylic)

Mary Pearce is a linguistics consultant, specializing in phonology and orthography, who works with a faith-based charity. She has spent many years researching, teaching and translating in Africa. She is now UK based as she has health challenges, including Parkinson's, that slow her down, but she is determined to keep smiling and enjoying life. She still finds enough energy to play music in a band and paint for pleasure. She particularly enjoys painting lions.

.

Mathias Jansson

160   *Crumbs of earth (poem)*

Mathias Jansson is a Swedish art critic and poet. He has contributed with poetry to different magazines and anthologies as Maintenant: A Journal of Contemporary Dada, Silver Birch Press and other publishers.
Homepage: http://mathiasjansson72.blogspot.se/
.

Michael Davies

31   *The Rise of the Crows (poem)*
34   *The Silence That Remained (poem)*

Michael Davies is a British film director, producer, and author, born in 1976 and proudly from Ellesmere Port. For 12 years, he worked in NHS operating theatres across multiple surgical specialties, supporting trauma and emergency teams on the front line. It was a career he deeply valued—until a life-changing accident caused a severe spinal injury that left him registered disabled.

In the aftermath of that fall, he faced a period of immense physical and emotional challenge. But it was during that time—when everything seemed uncertain—that he discovered a new purpose: storytelling.

What began as a coping mechanism soon became a passion. He wrote his first novel, *The Veil of Silence*—a psychological thriller exploring the fractured mind of a man battling internal demons. That book evolved into a screenplay and, eventually, a full feature film, which he directed in 2025. The film stars Simon Hall, Angel Nichole Bradford, Lisa Riesner, and Sarah Wingfield, and remains a faithful, chilling adaptation of the original story.

Building on that success, he penned *The Fractured Veil*—a companion piece and spiritual prequel. It dives into the mind of a single mother descending into madness, whose final connection is with the same man from The Veil of Silence. This interconnected narrative forms the foundation of a planned trilogy, with a third instalment, *Unraveling the Veil*, currently in development.

Since transitioning to film, he's directed and produced over 30 films, with a deep focus on psychological themes, identity, and human complexity. His work is known for leaving audiences unsettled, reflective, and eager to rewatch—always questioning what's real.

In addition to writing and directing, he's the founder of Dark Quill Films, through which he supports bold, independent cinema rooted in emotion and meaning. He's collaborated with countless talented actors, crew members, and fellow creatives, and is passionate about using storytelling as a way to connect, heal, and provoke thought. What began from tragedy has become a mission: to tell stories that matter—and to never stop creating, no matter the obstacles.
.

Morgan Rye

63   Cult of the Badger (painting)

Morgan Rye is a published poet, author and artist who lives in Wiltshire. Her first novel, *Snowball Earth – Quinn*, published by Black Eyes Publishing UK, is a coming-of-age story which explores the division between societal values and beliefs. Her second novel in the Snowball Earth series will be published in 2026.

Morgan has exhibited in art events and galleries across the southwest UK for many years, and is currently working on a series of portraits for the Swindon and Preservation Society. Examples of her poetry can be found in the online indie magazine, Steel Jackdaw and The Traveler 2021 and 2022.

.

Natasha Gauthier

59   The Birds are Hungry (poem)

Natasha Gauthier is a Canadian writer living in Cardiff. She is the 2025 New Welsh Review/Borzello Trust Prize winner, and also placed first in the 2024/25 Poetry Wales Awards. She placed runner-up and highly commended in the 2025 Wirral Poetry Festival competition, and is nominated for a 2025 Forward Prize (best single poem). Natasha has been published in Poetry Wales, New Welsh Review, Scintilla, Ink Sweat and Tears and Black Iris, among others. She was supported by Literature Wales' development programme, Representing Wales, in 2024-25, Natasha runs the Tiger Bay Poetry reading series.

.

Neil Windsor

81   Post Apocalyptical Zebra Crossing (artwork) ~ also used on back cover
83   Post Apocalypse Disco Badgers (poem)
85   Post Apocalyptical Disco Badgers (artwork)
87   Getting rid of old pizza boxes (artwork)

Neil Windsor is a writer, selling artist and poet from Leeds. His writing has an off-kilter element with occasionally a childish sense of the ridiculous. He has performed his written work regularly around the city and had pieces accepted by The Hedgehog Press, Dreamcatcher magazine, the 2024 Leeds Poetry Festival anthology and most recently three pieces published in the Blooming Under Bradford Skies anthology.

Neil also writes and illustrates short stories for children, some of which are available on Amazon / Kindle.

His art, in acrylic, pen and watercolour often reflects his northern roots. There is a strong sense of purpose and place in all he creates.

Neil works from his home studio and has surrounded himself with an eclectic array of objects accumulated over the years, touchstones of a life well lived and a source of constant inspiration.

naw21@aol.com #neilwindsorart

.

Nigel Kent

165   Dear Former Tennant (poem)

Nigel Kent is a Pushcart Prize nominated poet and reviewer who lives in rural Worcestershire. He is an active member of the Open University Poetry Society, managing its website and occasionally editing its workshop magazine.

He is the author of three collections *Fall, Unmuted* and *Saudade,* three pamphlets, *Sent, Benchwarmers* and *Psychopathogen* and two poetry conversations with Sarah Thomson, *Thinking You Home* and *A Hostile Environment*. All published by Hedgehog Poetry Press.

.

Peter Devonald

110   Dragons Sleep with One Eye Open (Tanka)
111   The Eyes of The Dragon (poem)
112   The Leftovers (poem)

Peter Devonald is a multi-award-winning poet/ screenwriter, published in over a hundred journals including five Broken Spine anthologies, Alchemy Spoon, London Grip, Dreich and Door Is A Jar. Winner Broken Spine's Reader's Choice Award 2025, Loft Books Best Poem 2025, Waltham Forest, Heart Of Heatons 2023 & 2021, joint winner FofHCS, runner-up Shelley Memorial and N2tS 2024. Finalist Tickled Pink ekphrastic, commended Bermondsey and Beyond 2025, Hippocrates and Passionfruit Review, shortlisted OxCanalFest 2024, Saveas & Allingham 2023. Nominated Forward Prize, two BestOfNet and Poet in Residence Haus-a-Rest. 50+ film awards, former senior judge/ mentor Peter Ustinov Awards (iemmys) and Children's Bafta nominated.

Facebook: @pdevonald
BSky: @pdevonald.bsky.social
Instagram: @peterdevonald
Twitter/X: petedevonald

.

Peter Lay

7	PREFACE
65	Rise of the Badger (poem)

Peter Lay is a writer and publisher living in Gloucester, although Peter doesn't consider himself a writer, more as someone who occasionally writes. Peter founded Black Eyes Publishing UK in January 2018, with Josephine Lay as editor. Black Eyes has published individual poets and writers as well as various anthologies.

From January 2020 till December 2022 Peter was the Events Co-Ordinator for the Gloucestershire Poetry Society (GPS), working alongside Josephine Lay who was at that time the GPS Director of Ops. and Jason Conway who was the design and tech director.

From 2020 – 2024 Peter ran the annual open poetry competition for the GPS, all profits going to help fund that society.

In a previous existence Peter was the Manager (Redbootsman) of the scrub-metal band, *'Oblivion'*, the rock band, *'AC-151'*, and the Dutch rock band, *'Journeys End'*. And before that he was a Youth Arts Worker in Gloucestershire and a Youth & Community Worker in London.

Peter has published 2 poetry collections under Black Eyes; *Becoming Naked (2023), Still Tilting at Windmills (2019)* and *Such Strange Philosophies, self-published (2016)*. Also, *Yellow Over the Mountain (2018)* a collaboration with Zaiming Wang – a philosophical conversation on a mythical journey of life, love and art; written in both English and Chinese. *Another Friday (2019)* poetry collection also with Zaiming Wang.

·

Rebecca Clifford

5	Quote from 'My mother was a Whizz in the kitchen'

Rebecca Clifford's poetry and prose has been widely published at home (Ontario, Canada) and in international anthologies and e-zines She lives rurally, near a watershed, gardens with a backhoe, and plants as many sunflowers as the ground will hold

Rhianna Levi

21   *These Facilities Are Not Exceptional (poem)*

Rhianna Levi is a writer, teacher, and bookshop social media manager. She is a former Worcestershire Poet Laureate, being published in numerous anthologies, literary magazines, and books. Rhianna has several degrees and as a writer and holistic educator, her work empathises the complexity of humanity and existentialism.

Facebook: Rhianna Levi
Instagram: @RhiannaLevi98
Threads: Rhiannalevi98
BlueSky: Rhiannalevi98

Rhian Thomas

5   *Quote from 'The Grandmother Hypothesis'*
41   *The Commons (poem)*

Rhian Thomas grew up on Anglesey where poetry was hard to avoid. After a couple of decades in London she can now see Wales across the Severn from her adopted home town of Stroud.

She has been shortlisted for the Laurie Lee Prize, the New Welsh Review's Borzello Prize, and published in journals including Poetry Wales and Ink Sweat and Tears. She was the joint winner of the Black Eyes Publishing and Gloucestershire Poetry Society's 2024 Friendship Prize.

Richard Catlin

127   *Microbial Soup (poem)*

Richard Catlin is 64, and lives in Stroud Gloucestershire. He has been writing since 2000 having attended a writing class for 4 years and occasional writing groups. He currently goes to Adam Horovitz's poetry workshop. He has not had anything published yet but has read some poetry at local events and written a few songs for a friend's band.

Roger Turner

22	*After the concert (poem)*
23	*Pellitory of the Wall (poem)*

Roger Turner worked as an architect and a garden designer, and has designed award-winning gardens at the Chelsea Flower Show and the Garden Festival in South Wales. He is the author of five non-fiction books, on landscape history and garden design and plants.

His poetry has been published in more than 100 reputable magazines. He is currently chairman of Cheltenham Poetry Society.

U.A.Fanthorpe said of his poetry: 'Turner is that not-often-found thing, a master of the sentence. His are like Hopkins' weeds, 'long, lovely and lush'.

•

Rosie Barrett

151	*The Planet's Doing Just Fine Thank You (poem)*

Rosie Barrett has lived in South Devon for more than half her life (just!). She's been writing on and off for as long as she can remember.

Shortlisted for the Bridport Prize she's had work published from Tasmania to the Mississippi as well as the UK. She likes Open Mics, telling stories, and is working towards a slim volume of her own.

•

Ruth Schreiber

95	*Old House in Moon Forest (watercolour)*

Ruth Schreiber is a multimedia artist whose subjects include landscapes and still life, the body and ageing, life cycle ceremony and practice, death and memory.

Ruth produces photography and installation pieces, paintings and sculpture, ceramics and video art. Her work can be found in private collections on three continents, and in several public collections. She is also a published poet and a volunteer docent at the Israel Museum Jerusalem.
www.ruthschreiber.com

•

Shashi Kadapa

122   It is the New Normal (flash-fiction)

Shashi Kadapa "Based in Pune India, Shashi Kadapa is the managing editor of ActiveMuse, a journal of literature. An engineer/ MBA, his stories across multiple genres are published in more than 45 US and UK anthologies. Winner of IHRAF, NY short story prize, he is nominated thrice for the Pushcart award.
His works: http://www.activemuse.org/Shashi/Shashi_Pubs.html"
.

Simon Alderwick

42   Elegy for a Parasite (poem)

Simon Alderwick lives in Oxford. His poetry has appeared in Magma, Ink Sweat & Tears, Acropolis, Frogmore Papers, Anthropocene, Berlin Lit and elsewhere. His pamphlet, *ways to say we're not alone*, is available from Broken Sleep Books.
.

Stephanie Carty

33   On not and always existing (flash-fiction)
39   Badgers on Coopers Hill (mixed media collage)

Stephanie Carty is a writer of all lengths in the U.K. with an interest in mixed media art. She has published two novellas-in-flash, two novels, a short fiction collection and two workbooks for writers. She has won Best Novella in the Saboteur Awards, Best Collection in the Eyelands Book Awards and was a recent winner in the Bath Novella-in-Flash Award.
linktr.ee/StephanieCarty
https://stephaniecarty.com/about/
X / Twitter - @tiredpsych
Bluesky - @stephaniecarty.bsky.social
Facebook – Stephanie Carty Author
Instagram - @stephaniecartyauthor
.

Stephen Chappell

48   When We're Gone (poem)

Stephen Chappell has had work published in Snakeskin, Flights of the Dragonfly, Ink Sweat and Tears and Dark Poets Club. He features in anthologies by Ledbury's HomEnd Poets, Malvern Spoken Word and The Hereabouts Poets. His first poetry collection *Portrait of the Artist with a Young Dog* came out in November 2024.
.

Stephen Littlejohn

114   Wegotu (poem)

Stephen Littlejohn has been writing for a while, previous awards include Alfred Bradley award for verse-plays run by the BBC, nomination for a Rhysling Award for best long science-fiction poem in the US. Has also self-published a novel under the name of Stephen Dee which is currently being serialised in the Yorkshire Press.
.

Stewart Carswell

75   Love song of the badger (poem)

Stewart Carswell grew up in the Forest of Dean. His poems have recently been published in Under the Radar, Finished Creatures, Ink Sweat & Tears, and The Storms Journal. His pamphlet "Knots and branches" was published in 2016, and his debut full-length collection "*Earthworks*" was published in 2021 by Indigo Dreams. Find out more at https://www.stewartcarswell.co.uk
.

Tom Edmonds

77   Above the Burrow (flash-fiction)

Tom Edmonds is a third-year Creative Writing student at the University of Gloucestershire, based in the UK. His fiction blends horror and transgression with humour and absurdity. Tom's work has been published in New Writing Vol. 13: Unbreakable Anthology (2024). He enjoys playing gigs, visiting galleries, and reading horror.
.

Tony Bradley

134   Unobserved (poem)

Tony Bradley now retired, Tony enjoyed a long career in both industry and teaching. Poetry is a lifelong interest, but he began writing poetry twenty years ago. Nature poetry is an area of interest, in particular the poetry of John Clare. He has also run an online poetry group for many years.
.

Trevor Valentine

129   Timber Cries (poem)

Trevor Valentine - singer-songwriter who uses poetry as a foundation for his songs. Based now in the Forest of Dean. Living in a former monastery, peace and quiet is now in abundance, which helps in writing his musical, due for completion in 2026. Poetry is the foundation of good songwriting. Music is what feelings sound like.

Trevor is currently working on a collection of his poems and songs with 'Red Balcony Press', an imprint of 'Black Eyes'.

·

Vidar Nordli-Mathisen

147   Orca (photo) ~ Licensed under the Unsplash License

Vidar Nordli-Mathisen is from Norway. Worked for many years as a professional photographer back in the day. Picked up photography for fun some 20 years ago when digital started to get serious. https://vidarnm.smugmug.com/

www.ingramcontent.com/pod-product-compliance
Lightning Source LLC
Chambersburg PA
CBHW051405070526
44584CB00023B/3306